Nuclear Engineering and Sciences Division

Core Programming Topic at the 2012 AIChE Annual Meeting

Pittsburgh, Pennsylvania, USA
28 October - 2 November 2012

ISBN: 978-1-62276-736-6

Printed from e-media with permission by:

Curran Associates, Inc.
57 Morehouse Lane
Red Hook, NY 12571

Some format issues inherent in the e-media version may also appear in this print version.

Printed by Curran Associates, Inc. (2013)

For permission requests, please contact AIChE
at the address below.

AIChE
3 Park Avenue
New York, NY 10016-5991

Phone: (203) 702-7660
Fax: (203) 775-5177

www.aiche.org

Additional copies of this publication are available from:

Curran Associates, Inc.
57 Morehouse Lane
Red Hook, NY 12571 USA
Phone: 845-758-0400
Fax: 845-758-2634
Email: curran@proceedings.com
Web: www.proceedings.com

TABLE OF CONTENTS

(205a) Rare Earth Elements: A US User's Perspective of Issues and Initiatives

Peter Dent, Electron Energy Corporation, Landisville, PA

Strategic and critical mineral supply chains are nationally recognized as being important to the US manufacturing industry, energy supply, and national defense. Currently, China dominates the rare earth market and will continue to do so even after new non-Chinese sources of supply come on line. The17 rare earth elements play a crucial role in international supply chains. The largest US industrial consumers by tonnage are catalyst producers who use rare earths for the cracking of crude petroleum and automotive catalytic converters for treating engine exhaust. Growing demand for alternative energy technologies which utilize rare earths include wind power, hybrid electric vehicles, advanced batteries, fluorescent lighting, and efficient permanent magnet motors and generators. To secure the future of these industries, we must address the challenges associated with supply risks due to their location, vulnerability to disruptions, and lack of suitable substitutes. This presentation will discuss the issues and initiatives in supply and demand in the complete rare earth supply chain and offer a user's perspective on navigating through this uncertain environment.

(205b) The Role of a Transformed National Defense Stockpile in U.S. Strategic Materials Security

Richard A. Lowden, Materials Science and Technology Division, Oak Ridge National Laboratory, Oak Ridge, TN

The United States established a stockpile of strategic materials in 1939 beginning with a handful of ores and minerals which with time was expanded to include additional minerals, metals, chemicals, and other processed materials. The purpose of the stockpile was to preclude a dangerous and costly dependence by the United States upon foreign sources for supplies of scarce but important materials in times of national emergency. Although emphasis has traditionally been on the acquisition and retention of certain strategic and critical materials, the legislation includes provisions that encourage the conservation and development of resources and materials. The authors of the Stockpiling Act realized the value of recycling, improved separation techniques, and alternative methods for refining or processing of materials thus included language to address these issues. Since 1993 the Stockpile has been downsizing, selling off most of its assets, however, recent changes in the global marketplace and increasing risks associated with the supply of important materials such as the rare earths, interest in the stockpile has been renewed. The Stockpile is in the process of being transformed in a more flexible and dynamic strategic materials risk management (security) program with much broader scope and applicability with respect to the entire supply chains for important materials.

(205c) Material Sustainability At General Electric

Anthony Ku and Steven Duclos, GE Global Research, Niskayuna, NY

GE Global Research, working with our GE business partners, has developed a comprehensive strategy to mitigate risks in the availability and sustainability of raw materials that are critical to high technology products. Many such products contain rare elements with limited availability due either to geopolitical constraints or the fact that they are refined as by-products of limited growth commodity metals. We have developed a methodology to assess such risks and prioritize technical solutions. These solutions include the development of advanced manufacturing techniques that minimize waste, recycling technologies to recover materials lost during manufacture or at end of life, and materials engineering that enables substitutions resulting in significant reduction in the use of at-risk elements without sacrificing material properties. Recent examples include our development of superalloys with 50% and 100% less rhenium, and magnet materials with reduced rare earth content. This presentation will discuss GE Global Research's work in this area.

(205d) The Issues and Opportunities Associated with Recycle and Reuse of Strategic and Critical Materials

Eric S. Peterson, Chemical Sciences, Idaho National Laboratory, Idaho Falls, ID

Strategic and critical mineral supply chains are nationally recognized as being important to the US manufacturing industry, energy supply, and national defense. For example, the largest industrial consumers of rare earths in the U.S. are catalyst manufacturers, representing almost 30% of the domestic market by volume in 2008 (40 to 45% in the previous five years). Rare earth elements are also essential to the production of catalysts for the cracking of crude petroleum and of automotive catalytic converters for treating engine exhaust and catalyst that contain rare earth elements influence the production of a diverse range of end user products from aircraft windows to dentures, plastic food packaging, etc. Many facets of the strategic and critical materials supply chain involve "chemistry" or "chemical engineering" from novel digestion and separation processes to reduction and refining to utilization in "old" and "new" products and technologies as well as recycle and reuse. This presentation discusses the issues and opportunities associated with recycle and reuse of critical materials.

(205e) Roundtable Discussion On Critical Materials Supply Chain and Sustainability

Sharon Robinson, Oak Ridge National Laboratory, Oak Ridge, TN

Strategic and critical mineral supply chains are nationally recognized as being important to the US manufacturing industry, energy supply, and national defense. Many facets of the strategic and critical materials supply chain involve chemistry and chemical engineering from novel digestion and separation processes to reduction and refining to utilization in "old" and "new" products and technologies as well as recycle and reuse. After a series of papers from leading experts in the field from both government and industry, the presenters will hold a round-table discussion to identify the issues and opportunities associated with the critical materials supply and sustainability.

Dependence of Hydrogen Adsorption on the Structural Properties of Carbons

Ali Qajar, Maryam Peer, Ramakrishnan Rajagopalan, Henry C Foley

Abstract:

Hydrogen adsorption properties of a series of polymer-derived, biomass derived and pyrolytic carbons are compared. PFA-derived carbons are synthesized and used the polymer derived carbons. The pyrolytics carbons are synthesized by using PEG-diacid modified PFA-derived carbon and ordered mesoporous carbons as substrates. Graphite was used as a reference to compare the importance of edge effects and interlayer spacing on the hydrogen adsorption properties. Boron as a heteroatom is doped into the pyrolytic carbon structure with no exposure to oxygen. Benzene cracking is used as a basis to compare the effect of heteroatoms and narrowing poresize of the carbons. Hydrogen adsorption data are collected at two pressure scales of up to 25 bar and 100 bar at room temperature. Presence of any chemistry between H_2 and carbons surfaces is examined by change in the temperature and sequential adsorption of CO_2, O_2 and H_2. Also Pt particles are embedded into the carbons to form carbon based catalysts. Presence of Pt is showing slight improvement in the adsorption uptake. However, a significant contribution to the H_2 adsorption is observed to be from the pore size of the carbon.

(259b) Carbon Assisted Hydrogenation and Dehydrogenation of Lithium Borohydrides

Mayfair C. Kung and Harold H. Kung, Chemical and Biological Engineering, Northwestern University, Evanston, IL

Application of complex metal hydrides as hydrogen storage materials is hampered by the slow hydrogenation and dehydrogenation kinetics. The presence of high surface area carbon greatly accelerated the dehydrogenation and hydrogenation kinetics of $LiBH_4$ even when the two components are present only as physical mixtures. Different synthesis procedures were explored to increase the contact between the two components. Dehydrogenation kinetics was accelerated for a carbon –$LiBH_4$ composite synthesized by liquid phase infusion of the carbon pores with 2M solution of $LiBH_4$, when compared to the physical mixture. After dehydrogenation, the sample was subjected to a rehydrogenation treatment of 390 °C and 14 atmospheres of H_2. H_2 uptake was observed even under such a mild condition, although only partial reversibility was observed after one hour of hydrogenation. The sample after the second dehydrogenation was rehydrogenated to about the same extent as the first rehydrogenation. The ability to effect rehydrogenation under mild conditions and short duration was attributed to a nano-confinement effect. Ni and Co were added to the carbon and tested as metal catalytic components to further accelerate the hydrogenation and dehydrogenation kinetics.

(259c) Improved Hydrogen Release From Magnesium Borohydride with Additives

Ahmad Al-Kukhun, Hyun Tae Hwang and Arvind Varma, School of Chemical Engineering, Purdue University, West Lafayette, IN

The dehydrogenation properties of magnesium borohydride ($Mg(BH_4)_2$
was found that certain additives can improve the dehydrogenation properties of magnesium borohydride significantly. While neat $Mg(BH_4)_2$ starts to release hydrogen > 270°C, $Mg(BH_4)_2$
release ~ 75°C, as confirmed by mass spectrometry and thermogravimetry. The hydrogen yield of $Mg(BH_4)_2$
presence of the additive was 3.7, 7.4, 10.0, 11.4 wt% for 150, 250, 300 and 350°C, respectively. Using pXRD, we confirmed that the final crystalline product at 300°C from $Mg(BH_4)_2$ with the additive was Mg, while it was MgH_2 for neat $Mg(BH_4)_2$. Solid state ^{11}B NMR analysis of $Mg(BH_4)_2$ with the additive at 300°C showed high selectivity towards the formation of $Mg(B_{12}H_{12})$ as intermediate, while neat $Mg(BH_4)_2$ showed β-$Mg(BH_4)_2$
well as some $Mg(B_{12}H_{12})$. Our results demonstrate that the additive is a promising one to provide high hydrogen yield values from $Mg(BH_4)_2$ at moderate temperatures < 300°C, suitable for vehicle applications.

Catalytic Effect of Boric Acid on

Thermal Dehydrogeantion of Ammonia Borane

Hyun Tae Hwang, Ahmad Al–Kukhun, and Arvind Varma

School of Chemical Engineering, Purdue University,
480 Stadium Mall Drive
West Lafayette, IN 47907, USA

ABSTRACT

Ammonia borane (AB, NH_3BH_3) is a promising hydrogen storage material for use in proton exchange membrane (PEM) fuel cell applications. In this study, the effect of boric acid on AB dehydrogenation was investigated. Our study shows that boric acid is a promising additive to decrease onset temperature as well as to increase hydrogen release kinetics for AB thermolysis. Using boric acid, we obtained high H_2 yield (11.5 wt% overall H_2 yield, 2.23 H_2 equivalent) at 85°C, PEM fuel cell operating temperatures, along with rapid kinetics. In addition, only trace amount of NH_3 (20-30 ppm) was detected in the gaseous product. Studies to identify the catalytic mechanism of boric acid were also conducted. The spent AB solid product was found to be polyborazylene-like species, which can be efficiently regenerated using currently demonstrated methods. The results indicate that our proposed method is promising for hydrogen storage, and could be used in PEM fuel cell based vehicles.

(259e) Active Role of Carbon Dioxide in the Dehydrogenation of B-N-H Compounds

Ran Xiong[1], Junshe Zhang[2,3] and Jae W. Lee[2], (1)Chemical Engineering, CCNY, New York, NY, (2)Department of Chemical Engineering, The City College of New York, New York, NY, (3)Chemical and Biomolecular, KAIST, Daejeon, South Korea

B-N-H compounds including ammonia borane (AB, NH_3BH_3) and polymeric aminoborane (PAB, (H_2BNH_2) n) are promising on-board hydrogen storage materials. To satisfy the DOE 2017 target (5.5 wt% H_2 based on the total system), about 11 wt% H_2 should be usable from a storage material. Thus, the research challenge is to propose a facile and economical method to extract 11 wt% H_2 from the starting material of AB at the working temperature of fuel cell (80 – 90 °C). The present study has found that gaseous CO_2 environments can enhance the dehydrogenation of AB and PAB. After AB is pre-treated at 70 °C and with 1.38 MPa of carbon dioxide (CO_2) for 38 min, the rapid hydrogen release of 1.5 moles H_2 per mol of AB was achieved within 1 hour at 85°C. Furthermore, when PAB, the spent fuel of AB with 0.91 equivalent hydrogen release, was exposed to 2.07Mpa of CO_2 at 85°C, the rapid hydrogen desorption of 0.84 mol H_2 per mol of AB was observed. The total hydrogen release starting from AB is 11.4 wt% based on an initial mass of AB. For both cases, the degree of dehydrogenation is confirmed by ATR-FTIR spectroscopic and elemental analyses of the thermolytic solid products. Based on the IR, Raman, and elemental analyses of the solid products, we will address a possible mechanism for the enhanced dehydrogenation of AB and PAB in CO_2 environments.

(259f) Hydrogen Storage and Carbon Dioxide Sequestration in Tbaf Semi-Clathrate Hydrates: Kinetics and Evolution of Hydrate-Phase Composition by in Situ Raman Spectroscopy

Alondra Torres Trueba, Process & Energy, Delft University of Technology, Delft, Netherlands

Hydrogen storage and Carbon Dioxide sequestration in TBAF Semi-Clathrate Hydrates: Kinetics and Evolution of Hydrate-Phase Composition by *in situ* Raman Spectroscopy

Alondra Torres Trueba,[1] Ivona R. Radoviæ,[1] John F. Zevenbergen,[2]

Maaike C. Kroon,[3] Cor J. Peters[1,3,4*]

[1]Department of Process & Energy, Delft University of Technology, Delft, Netherlands

[2]TNO Defense, Security and Safety, Lange Kleiweg 137, 2288 GJ Rijswijk, Netherlands

[3] Department of Chemical Engineering and Chemistry, Eindhoven University of Technology, Eindhoven, Netherlands

[4]Chemical Engineering Program, Petroleum Institute, Abu Dhabi, United Arab Emirates

ABSTRACT

Carbon dioxide (CO_2) represents almost one third of the emissions from the combustion of fossil fuels additionally, CO_2 has been identified as the mayor contributor of global warming. Hydrogen (H_2), on the other hand, due to its properties is considered a promising energy carrier. Clathrate hydrates are a potential alternative for CO_2 separation, capture and storage (CCS) as well as a medium for H_2 storage. For instance, clathrate hydrate technologies are less energy demanding and environmentally friendly than other technologies available. However the main disadvantage of clathrate hydrates is the high pressure required for their formation. Recently, tetra-n-butyl ammonium fluoride (TBAF) has been used as an additive to form H_2-TBAF and CO_2-TBAF semi-clathrate hydrates with water, which are stable at much more practical conditions (atmospheric pressure and close to room temperature). Moreover, TBAF is non-volatile which means there is no contamination of the gas phase. These favorable conditions make these two semi-clathrate hydrates particularly attractive for hydrate based technologies.

In order to investigate the potential of H_2-TBAF and CO_2-TBAF semi-clathrate hydrates for H_2 storage and CO_2 sequestration, kinetic data were obtained. The influence of pressure, solute concentration and formation rate was elucidated. Coupled with the kinetic results, the Raman spectra studying the structural changes that take place during formation and dissociation of the H_2-TBAF and CO_2-TBAF semi-hydrates were recorded *in situ* by Raman spectroscopy. To the best of our knowledge, these are the first results of time-dependent kinetic conditions of H_2-TBAF and CO_2-TBAF semi-clathrate hydrates using *in situ* Raman spectroscopy in an aqueous solution. Results will be presented at the conference.

* Corresponding author. Tel.: +31-15-2782660; Fax.: +31-15-2786975; E-mail: cpeters@pi.ac.ae

(347a) Investigation of Phase Stability of Magnesium Alanate for Hydrogen Storage From First Principles

Dong-Hee Lim, Energy Resources Engineering, Stanford University, Stanford, CA, Tim Mueller, Materials Science and Engineering, Johns Hopkins University, Baltimore, MD and Jennifer Wilcox, Department of Energy Resources Engineering, Stanford University, Stanford, CA

Complex hydrides including alanates ($[AlH_4]^-$) have recently gained attention as alternative hydrogen storage materials. Many of these materials have been known to release hydrogen upon contact with water; however, the hydrolysis reactions are highly irreversible, a process known as "one-pass" hydrogen storage. Nanostructuring and nanocatalysis have been accepted as promising methods to overcome the irreversible hydrogenation process. Thus, predicting which phases may be more stable as a function of nanoparticle size may contribute to nanostructuring complex hydrides for hydrogen storage applications. We have employed density functional theory (DFT) using the projector-augmented wave (PAW) method within the generalized gradient approximation (GGA) to calculate relatively smaller nanoparticles of magnesium alanate ($Mg(AlH_4)_2$) ranging from 1 to 2 nm. Based upon these results, cluster expansion and Monte Carlo simulation methods were developed to predict the phase stabilities of 2-10 nm $Mg(AlH_4)_2$ nanoparticles. Our calculations provide phase stability diagrams of $Mg(AlH_4)_2$ nanoparticles as a function of particle size and temperature. This study may help identify how the relative stability of different compounds ($Mg(AlH_4)_2$, MgH_2, Al, and H_2) evolves as a function of nanoparticle size and temperature, which will facilitate experimental studies to determine the thermodynamically favored reaction pathways for the hydrogenation/dehydrogenation processes of $Mg(AlH_4)_2$.

(347b) Role of Dopants in Dehydrogenation of NaMgH3 Core-Shell Nanostructures

Fernando A. Soto, Louisiana Tech University, Ruston, LA and Daniela S. Mainardi, Chemical engineering, Louisiana Tech University, Ruston, LA

Role of Dopants in Dehydrogenation of $NaMgH_3$ Core-Shell Nanoclusters

Authors: Fernando Soto[1], Daniela S. Mainardi[2]

[1]Institute for Micro-manufacturing, Louisiana Tech University

[2]Institute for Micro-manufacturing, Chemical Engineering, Louisiana Tech University

Abstract

$NaMgH_3$ is an attractive hydrogen storage material due to its 6 wt % hydrogen capacity and its hydrogenation/dehydrogenation reversibility. However, more fundamental understanding of the role of dopants to potentially improve the thermodynamics and kinetics of this hydride system is needed. Previous results show that clusters of hydrogen storage materials have a much lower desorption energy that their bulk counterparts; hence, they enable hydrogen desorption at lower temperatures[2].

In our research group, the synergistic effects of 3*d* transition-metal co-dopants on the dehydrogenation kinetics and thermodynamics of $NaMgH_3$ have been studied using first-principles calculations based on density functional theory. Results indicate that co-dopants introduced at the (001) $NaMgH_3$ surface model improves the reaction kinetics by lowering the Gibbs Free-Energy barriers. This indicates that $NaMgH_3$ is suitable for nano-structured modification by 3*d*- transition metal dopants.

In this paper, 3*d* transition-metal doped $NaMgH_3$ nanoclusters with core-shell features are investigated using density functional theory to systematically explore their segregation energy, cohesive energy, substitutional doping energy, and hydrogen desorption energy to determine 3*d* transition metal core-shell preference. These calculations provide a general trend for nano-structured manipulation of hydride systems geared towards reducing the current high temperature hydrogen desorption barrier, and provide a more detailed understanding of dopant behavior.

(347c) Adsorption Based Hydrogen Storage System; Multiscale Approach and Mathematical Model

Agnieszka Truszkowska, Oregon State University, Corvallis, OR, Christopher Loeb, Oregon State University and Goran Jovanovic, School of Chemical, Biological, and Environmental Engineering, Oregon State University, Corvallis, OR

Adsorption Based Hydrogen Storage System; Multiscale Approach and Mathematical Model

Agnieszka Truszkowska[1], Christopher Loeb[1] and Goran Jovanovic[1]

[1]Oregon State University, Department of Chemical, Biological, and Environmental Engineering, Corvallis, OR 97330, USA

Emails: truszkoa@onid.orst.edu, loebc@onid.orst.edu, goran.jovanovic@oregonstate.edu

In an effort to reduce CO_2 emission arising from sources of current forms of transportation energy, hydrogen is sought as a potential replacement to major fossil fuels. However, deploying hydrogen into vehicles faces considerable challenges, one of which is efficient storage solution that would meet practical performance and cost criteria. Adsorption based storage systems are one of the promising storage solutions, which are currently under intensive development.

Adsorption based storage systems are typically made of materials like activated carbon and MOF-5; hence, they are considered a porous structure. These materials often meet gravimetric storage requirements, but their volumetric capacities become sufficient only at cryogenic conditions. Operation of hydrogen storage system in cryogenic region creates opportunities for many design solutions.

In this paper we propose a numerical tool based on ***multiscale mathematical modeling***, which supports experimental effort in performance characterization of cryogenic hydrogen adsorption storage system.

Mathematical model and corresponding numerical simulation techniques provide a tool capable of numerically describing hydrogen storage phenomena on an arbitrary number of structural levels in three dimensions.

Numerical package provides users with essential properties of porous hydrogen storage media: permeability, effective diffusivity and effective conductivity of the system. These mixed properties are based on material, process, and geometric characteristics of the storage system.

Results of numerical simulation enable a valid analysis of system performance at each scale and provide means for further optimization. We also propose the ***characteristic times analysis*** as one of the modeling approaches which informs design development of the hydrogen storage system.

In addition we provide validation of the mathematical model and the results of numerical simulation with ***experimental data collected*** on adsorption media (activated carbon and MOF-5) in our hydrogen storage laboratory.

(347d) Kinetics of Hydrolysis of Sodium Borohydride with Water Vapor

Lin Yu and Michael A Matthews, Chemical Engineering, University of South Carolina, Columbia, SC

Hydrolysis of sodium borohydride is one route for solid state hydrogen storage and release. Compared with the aqueous phase reaction, the hydrolysis of sodium borohydride with water vapor gives a faster reaction for hydrogen production. Prior studies on the water vapor hydrolysis from our group have revealed the physical mechanism for the release of hydrogen by water vapor. However, there are presently no data on the kinetics of the hydrogen evolution. We have developed a semi-batch tube reactor to study the hydrolysis of sodium borohydride with water vapor. The water vapor deliquesces to form a concentrated liquid phase that reacts to evolve hydrogen. It is found that $CoCl_2$, a solid catalyst, can significantly improve the hydrogen generation rate. The impact of this metal salt catalyst on the deliquescence and hydrolysis kinetics has been investigated. This presentation will be the first to report a quantitative study on the hydrogen generation kinetics of the steam phase hydrolysis. The hydrogen delivery rate for $NaBH_4$ reacting with water vapor has also been optimized.

(347e) Hydrogen Release Characterization and Simulation for Safe Design of Hydrogen Fueling Station

Jaedeuk Park[1], Eunjung Kim[2], Hyuckmyun Kwon[3], Chanho Tae[4] and Il Moon[1], (1)Chemical and Biomolecular Engineering, Yonsei University, Seoul, South Korea, (2)Gas Safety R&D, Korea Gas Safety Corporation, Siheung-si, Kyonggi-do, South Korea, (3)Center for chemical Plant Safety, Korea Occupational Safety and Health Agency, Incheon, South Korea, (4)GS Caltex Corporation, Seoul, South Korea

Hydrogen is a prominent fuel for a renewable and environment-friendly energy carrier. It can simultaneously reduce a country's reliance on foreign oil and considerably reduce greenhouse gases. Significant advances have been made in the use of hydrogen as a transportation fuel and a fuel for power generation. Hydrogen can be used in an internal combustion engine or a fuel cell to generate power. A number of hydrogen fueling stations are under construction to fuel hydrogen-driven vehicles. It would be essential to ensure the safety of hydrogen station equipment and operating procedure in order to prevent any leak and explosions of hydrogen: safe design of facilities at hydrogen fueling stations e.g. pressurized hydrogen leak from storage tanks. Several researches have centered on the behaviors of hydrogen ejecting out of a set of holes of pressurized storage tanks or pipes. Y. Lee et al. [1] have developed an effective operator training program providing the safety information and the public relations for the safer usage of hydrogen at hydrogen fueling stations. Some studies related to the safety of hydrogen stations have been concerned with the diffusion of leakage, explosion, deflagration or detonation of hydrogen and jet flames from hydrogen fueling stations [2].

This work focuses on the experiment and 3D simulation of hydrogen leak scenario cases at a hydrogen fueling station, given conditions of a set of pressures, 100, 200, 300, 400 bar and a set of hydrogen ejecting hole sizes, 0.5, 0.7, 1.0 mm. The physical points for probing hydrogen concentrations along with the centerline of leak jet stream are positioned at a set of horizontal distances away from the hydrogen cylinder; 1, 3, 5, 7 and 9 m. Hydrogen concentrations are measured from 5 samplers and collected from 5 probe points. In this work, the simulation is performed using FLACS because one of the primary aims of this study is to identify safe distances in case of hydrogen explosion and FLACS, a CFD tool designed to simulate behaviors of both leak and explosion, is used to calculate the concentration profile of high pressure hydrogen in released jet stream. The simulation results are validated with experimental data and compared with those of previous study [3]. The simulation is based on real 3D geometrical configuration of a hydrogen fueling station that is being commercially operated in Korea. The simulation results are validated with hydrogen jet experimental data to examine the diffusion behavior of leak hydrogen jet stream.

The safety distance is defined as a distance between a major facility and building where people reside. Each nation has its own standard for the safety distances. It is defined as a distance at which the allowable burst pressure of 0.1226 bar is considered on the premise that it will not cause any human damage to the surroundings of the business or any property loss except minor ones. To identify the safety distances in hydrogen fueling stations, explosion scenario-based dynamic simulation is made, given a layout of facilities comprising the station. Hydrogen blast pressure profiles are calculated in case of explosion using FLACS, which provides information on the distribution characteristics of blast pressure through geometric facility configurations, and the directionality and development of explosion. 3D Simulation results of scenarios are shown in Fig. 1. Finally, a set of marginal safe configurations of fueling facility system are presented, together with an analysis of distribution characteristics of blast pressure, directionality of explosion. This work makes an important contribution to an enactment of official hydrogen safety distance in Korea being in the absence of such a standard as well as marginal hydrogen safety design for hydrogen fueling stations, enabling to keep a reasonable balance between less construction cost and high safety.

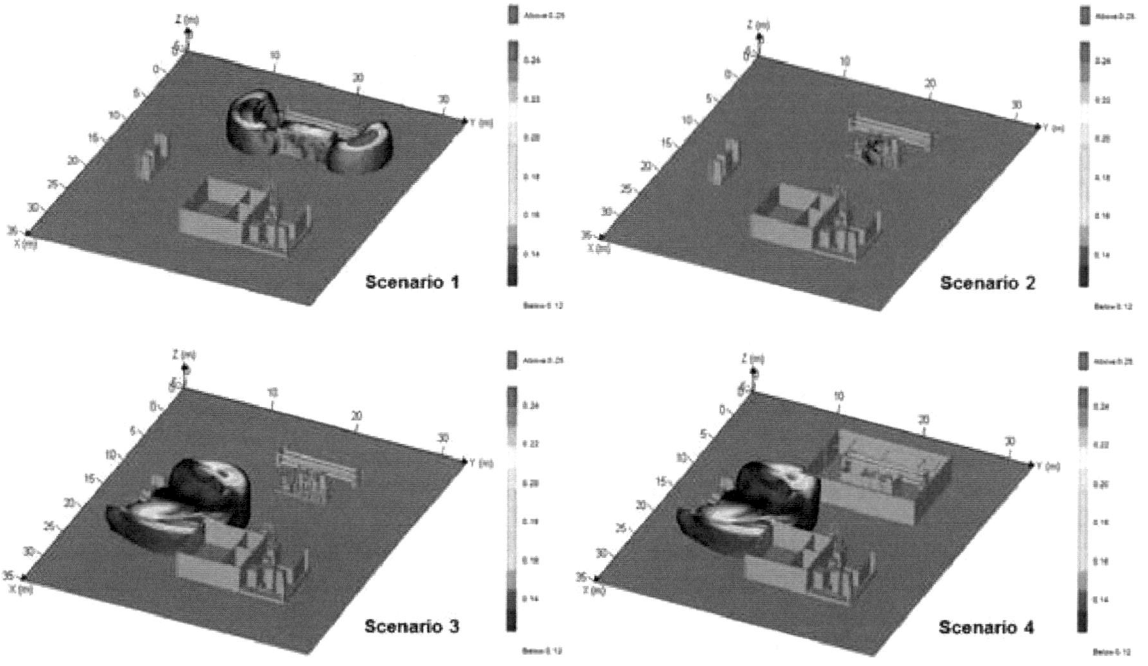

Fig. 1. 3D Simulation results of scenarios

REFERENCES

[1] Lee YH, Kim JK, Kim JH, Kim EJ, Kim YG, Moon I. Development of a web-based 3D virtual reality program for hydrogen station. Int J Hydrogen Energy 2010;35:2012-8.

[2] Zheng J, Bie H, Xu P, Liu P, Zhao Y, Chen H, Zhao L. Numerical simulation of high-pressure hydrogen jet flames during bonfire test. Int J Hydrogen Energy 2012;37:783-90.

[3] Kim EJ, Lee KW, Kim JS, Lee YH, Park JD, Moon I. Development of Korean hydrogen fueling station codes through risk analysis. Int J Hydrogen Energy 2011;36:13122-31.

(347f) Interaction of Hydrogen Molecules with Defective Graphene Sheets: An Ab Initio Study

Roghayyeh Lotfi and Yadollah Saboohi, Sharif University of Technology, Tehran, Iran

Hydrogen appears as one of the most promising energy carriers. Hydrogen has the advantage of being the cleanest fuel and as a result it can efficiently help limiting the greenhouse effect. However one of the major barriers for the large scale use of hydrogen, especially in the transportation sector, is hydrogen storage issue. There are four different technologies for storing hydrogen which are as follows: (1) Cryogenic liquid hydrogen (LH); (2) compressed gas storage; (3) metal hydride storage technology; (4) Physisorption methods. Physisorption of hydrogen in novel carbon nanostructures such as carbon nanotubes and graphene seems to be a safe and promising method for storage of hydrogen. But adsorption energy of pristine perfect carbon nanostructures is not strong enough for storing adequate amounts of hydrogen. On the other hand existence of defects can increase the hydrogen capacity, by reducing the barrier energy of dissociative hydrogen adsorption. Despite the research works have been performed so far on the adsorption of hydrogen on graphene, hydrogen interaction with different defective structures of graphene is still to be reviewed. In the present research work, ab initio calculations have been carried out to study the interaction of hydrogen with different defects in graphene. Also the barrier energies in defective graphene sheets are calculated and the results are compared with the defect-free graphene. The results indicate that defects reduce the barrier energy significantly. For instance, SW defect reduces the barrier about 41 percent relative to the perfect graphene, while 555777 and 585 defects decrease the barrier energy about 46 and 65 percent, respectively.

(400a) The Oxidation Kinetics of Two Step Solar Thermochemical H2O and CO2 Splitting with CeO2

Darwin Arifin, Chemical and Biological Engineering, University of Colorado Boulder, Boulder, CO, Anthony H. McDaniel, Sandia National Laboratories, Livermore, CA and Alan W. Weimer, Chemical and Biological Engineering, University of Colorado at Boulder, Boulder, CO

The two step solar thermochemical H_2O and CO_2 splitting with CeO_2 has been investigated. We observe that the thermal reduction process of CeO_2, to form the nonstoichiometric $CeO_{2-\delta}$, is fast for particles of order 5 μm diameters. Furthermore, we introduce a novel numerical model fitting method to implement solid state kinetics theory to study the oxidation process of $CeO_{2-\delta}$. The advantage of this method is that the kinetic parameters obtained are fitted over the complete conversion range and not limited to a specific range of reaction extent. We are unaware of this methodology used to analyze high temperature ceria kinetics elsewhere in the literature.

The objective of the kinetic analysis is to determine the intrinsic kinetic parameters of the oxidation process: energy of activation (E_a), reaction mechanism ($f(\alpha)$), and oxidant concentration dependence (γ). In this work, we determine that the oxidation of $CeO_{2-\delta}$ by H_2O and CO_2 is surface reaction limited in the range of $0 < d < 0.04$. The H_2O oxidation kinetics can be modeled very well; the kinetic parameter for the water splitting reaction is reported. CO_2 oxidation on the other hand, cannot be adequately described by a simple application of solid state kinetic theory because it fails to describe the complete temporal behavior of the CO evolution process.

(400b) Hydrogen Production Through a Modified Sulfur-Iodine Thermochemical Water Splitting Cycle in Hydrophobic Ionic Liquid

Kevin Caple, School of Chemical, Biological and Environmental Engineering, Oregon State University, Corvallis, OR

Hydrogen Production through a Modified Sulfur-Iodine Thermochemical Water Splitting Cycle in Hydrophobic Ionic Liquid

Kevin Caple and Alexandre Yokochi

One of the most pressing issues of the new millennium is the development of clean, simple, and renewable energy sources. Of the various process that have been studied, thermochemical water splitting is one that has been conceptualized and researched for over half a century, yet, at this point, none are commercially viable. These thermochemical water splitting cycles include the family of Sulfur cycles and metal/metal oxide couples.

Of the Sulfur based cycles, one of the simplest is the Sulfur-Iodine cycle, which uses sulfur dioxide, iodine, and a large excess of water to produce iodic acid and sulfuric acid, both of which can be further processed into hydrogen and oxygen gases, and regenerate the initial sulfur dioxide and iodine for reuse. Unfortunately, this process has been stalled due to difficulties separating the azeotropic iodic acid/water mixture.

We have developed an approach to avoid this issue in utilizing what was previously considered an undesired side reaction that produces hydrogen sulfide. It has been shown that hydrogen sulfide can be successfully steam reformed into hydrogen, as well as regenerate sulfur dioxide to complete this modified Sulfur-Iodine cycle. The series of reactions is as follows:

$I_2 + SO_2 + 2H_2O$ -> $2HI + H2_SO_4$ (Bunsen Reaction, Low Temp)

$8HI + H_2SO_4$ -> $H_2S + 4I_2 + 4H_2O$ (Side Reaction Producing H2S, Low Temp)

$H_2S + 2H_2O$ -> $3H_2 + SO_2$ (Steam Reformation, High Temp)

In our work we carry out the two low temperature reactions in an imidazolium based hydrophobic ionic liquid since both iodine and sulfur dioxide have solubilities in these. In our presentation, we will discuss recent experimental results of this reaction in ionic liquid, including kinetic parameters, the results of regenerating of the ionic liquid, and the extent of the cyclic reaction.

THERMALLY STABILIZED FERRITE NANOPARTICLES FOR HYDROGEN GENERATION FROM THERMOCHEMICAL WATER-SPLITTING REACTION

X. Pasala, V. S Amar, J. A. Puszynski, R. V. Shende
South Dakota School of Mines and Technology, SD, USA

Abstract

In this study, $NiFe_2O_4$ was synthesized using sol-gel technique in which salts of Ni and Fe were sonicated in ethanol until a visually clear solution was obtained. To this solution, propylene oxide was added to achieve the gel formation. As-prepared gel was dried at 100°C for 1 h and calcined upto 600°C at a ramp rate of 40°C/min. The calcined sample from the furnace was removed at 600°C and cooled down at room temperature in air. As-calcined $NiFe_2O_4$ powder was ultrasonically mixed with ZrO_2 nanoparticles as grain growth inhibitor. This powdered material was analyzed using powder x-ray diffraction (XRD), BET surface area analyzer and scanning (SEM) and transmission electron microscopy (TEM). As-prepared $NiFe_2O_4$ or ZrO_2 mixed $NiFe_2O_4$ powdered material was loaded in an Inconel tubular reactor to investigate H_2 generation from four consecutive thermochemical cycles where water-splitting and regeneration was performed at 900° and 1100°C, respectively.

Introduction

Solar assisted thermochemical water-splitting constitutes one of the promising options for the H_2 production [1-5]. Solar concentrating systems are able to provide the necessary temperatures to drive the two-step thermochemical water-splitting process [6], which utilizes redox reactions of metal oxides (MO) [1]. In the first step, the reduced MO is oxidized by taking O_2 from water producing H_2 via water-splitting reaction. Whereas in the second step, the oxidized MO is converted back to its reduced form by releasing O_2 during thermal reduction step [7].

Different redox materials such as Ni-ferrite, Zn-ferrite, Mn-ferrite, Co-ferrite, Ni-Mn-ferrite, Ni-Sn-ferrite, Ni-Zn-ferrite, ZnO/Zn, and FeO/Fe_2O_3 have been investigated for thermochemical H_2 production via water-splitting step. Among these materials, Ni-ferrite was reported to be the most promising material for H_2 production [ref]. Fresno et al.[6] investigated the H_2 generation ability of commercially available Ni-ferrite from Sigma Aldrich and reported 7.6 mL H_2/g·cycle in three consecutive thermochemical cycles where water-splitting and regeneration steps were carried out at 1000°C and 1300°C, respectively. Contrasting to this, Bhosale et al. [7] reported 31.63 mL H_2/g·cycle using sol-gel derived Ni-ferrite in four consecutive thermochemical cycles, where water-splitting and regeneration of the reacted ferrite was carried out at 900°C. Recently, our group has reported H_2 generation ability of several other un-doped and doped ferrites prepared using sol-gel technique [2,7,8].

Although, the sol-gel derived ferrites are capable of producing higher levels of H_2 at low to moderate temperatures, the H_2 generation ability was significantly decreased during multiple thermochemical cycles [2,7,8]. These observations were also reported by other investigators in case of ferrite materials synthesized using different methods such as solid state synthesis, aerosol spray pyrolysis and self-propagating high temperature synthesis (SHS). In general, the temperatures corresponding to the regeneration step is typically higher than water-splitting step making materials to undergo thermal cycling. As a result grain growth occurs due to sintering, which decrease specific surface area (SSA) and porosity. It has been observed by our group that both these factors decrease the overall H_2 volume in multiple thermochemical cycles. Therefore, it is critical to prevent the grain growth of redox materials to achieve steady hydrogen levels in multiple cycles. To address this issue, a grain growth inhibitor may be added to redox materials. Very recently, several grain growth inhibitors were mixed with Ni-ferrite using vortex mixing, which showed improvement in H_2 volume generation and inhibition of grain growth.

In this investigation, we report the synthesis of $NiFe_2O_4$ via sol-gel technique. As an attempt to improve the grain growth inhibition, $NiFe_2O_4/ZrO_2$ powdered mixture was prepared by adding 25 wt% ZrO_2 to the sol-gel derived $NiFe_2O_4$. The H_2 generation and regeneration abilities of both these powdered materials were investigated and compared by performing multiple thermochemical cycles.

Experimental

Preparation of Ni-ferrite/ ZrO_2 mixture

To synthesize Ni-ferrite, the chlorides of Ni and Fe were taken in stoichiometric weight ratios. Specifically, 10 gms of Ni and Fe precursors were added into 20 ml of the ethanol and sonicated for about 1 hr until visually clear solution was obtained. To this solution, 20 ml of propylene oxide (PO) was added to achieve gel formation. As-obtained gel was aged for 24 hr at room temperature, dried for about an hour at 100°C and calcined at 600°C at the rate of 40°C/min. The material was rapidly cool down to room temperature. As-calcined Ni-Ferrite powdered material was mixed with ZrO_2 nanoparticles (25 wt% based on Ni-Ferrite) and added into isopropyl alcohol (IPA) which was sonicated for about 2 hr to achieve homogeneous mixing. Well homogenized slurry or dispersion thus obtained was dried on a hot plate for 1 hr at about 100°C. Using this powdered mixture four consecutive thermochemical cycles were performed.

Characterization

The specific surface area of sol-gel derived $NiFe_2O_4$ and $NiFe_2O_4/ZrO_2$ was measured using BET surface area analyzer (Gemini II-2375 from Micromeritics). The phase composition was determined using Rigaku Ultima-Plus X-ray diffractometer (*CuKα* radiation, λ=1.5406 Å, 40 kV, 40 mA) provided with a graphite monochromator. The parameters such as 2θ, scanning speed and width of $10^\circ \leq 2\theta \leq 80^\circ$, 2° per minute and 0.020°, respectively, were used for the X-ray diffraction measurements. The morphology of the mixture was investigated using Zeiss Supra 40VP field-emission scanning electron microscopy (SEM) and Hitachi H-7000 FA transmission electron microscope (TEM).

Thermochemical water-splitting reaction set-up

The measured amount of $NiFe_2O_4/ZrO_2$ powdered mixture was packed in the middle of Inconel tubular reactor and packed with Raschig rings. The reactor was mounted inside a vertical split furnace. A horizontal tube furnace with a stainless steel tube was used to vaporize the water. The water from the reservoir was fed to the horizontal furnace using a metering pump at the flow rate of 1 mL/min. The N_2 at the flow rate of 100 ml/min was fed as a carrier gas. The exit gas stream from reactor was continuously monitored using an online H_2 sensor (H_2 Scan) and the data was continuously acquired using a computer. The water-splitting was performed at 900°C and the regeneration was carried out at 1100°C.

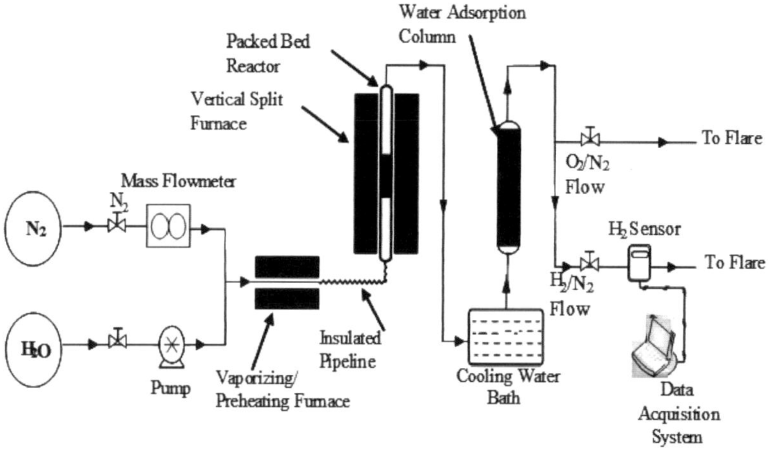

Figure 1 Thermochemical water-splitting reactor set-up for H_2 generation [2].

Results and Discussion

$NiFe_2O_4$ was synthesized using the sol-gel technique described in previous section. The calcined $NiFe_2O_4$ powdered material was untrasonically mixed with ZrO_2 nanoparticles (25 wt% based on $NiFe_2O_4$). The phase composition of the calcined $NiFe_2O_4$ and the powdered mixture of calcined $NiFe_2O_4$ and ZrO_2 nanoparticles was analyzed using x-ray diffractometer and the XRD patterns obtained are shown in Figure 2. The analysis of the 2θ peak positions revealed nominally phase pure material consistent with ICDD patterns.

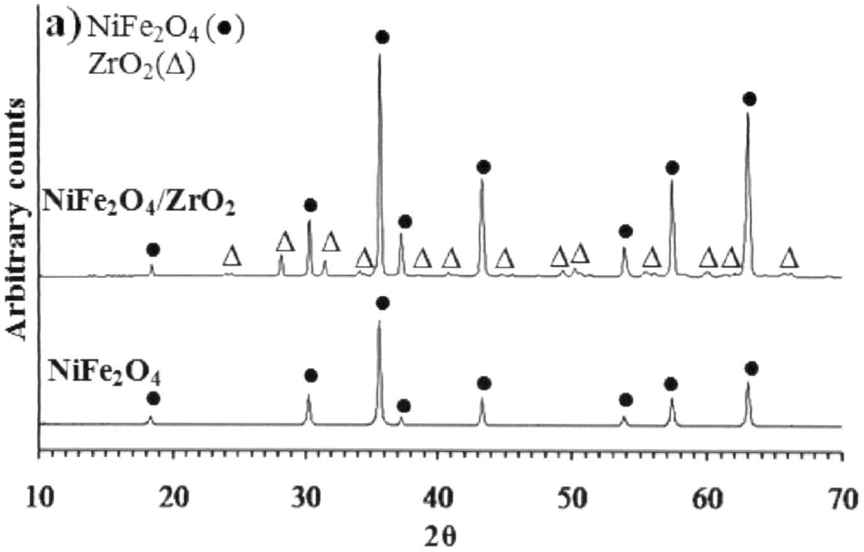

Figure 2 XRD patterns of reacted Ni-ferrite and Ni-ferrite/ZrO_2 powdered materials

BET specific surface area of both $NiFe_2O_4$ and $NiFe_2O_4/ZrO_2$ mixture was 35.55 and 33.70 m^2/g, respectively. Further, the morphology of both the powders was analyzed using TEM and the image obtained is shown in Figures 3a, which indicates nanosize particles. The SEM and EDX mapping of $NiFe_2O_4/ZrO_2$ material are shown in Figures 3b and 3c, respectively. The EDX mapping shows relatively better distribution of ZrO_2 nanoparticles in a powder mixture.

Figure 3 a) TEM, b) SEM and c) EDX mapping of $NiFe_2O_4/ZrO_2$ powdered mixture

The H_2 generation ability of sol-gel derived $NiFe_2O_4$ and $NiFe_2O_4/ZrO_2$ powdered mixture was investigated by performing four consecutive thermochemical cycles where water-splitting and regeneration steps were performed at 900°C and 1100°C, respectively. After conducting the water-splitting step, the material was regenerated at 1100°C for 2 h in N_2 environment and the reactor was again cooled down to 900°C to perform subsequent thermochemical cycles. The transient H_2 profiles obtained for sol-gel derived $NiFe_2O_4$ and $NiFe_2O_4/ZrO_2$ powdered materials during four thermochemical cycles are presented in Figure 4.

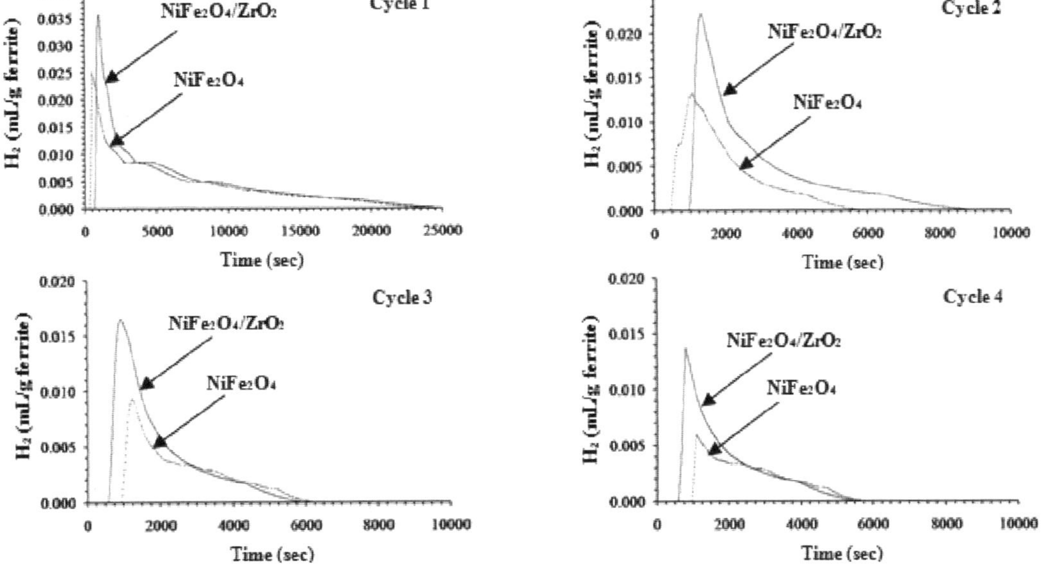

Figure 4 Transient H_2 profiles obtained during four thermochemical cycles where water-splitting was performed at 900°C using $NiFe_2O_4$ and $NiFe_2O_4/ZrO_2$ powdered mixture.

From the transient H_2 profiles obtained, the H_2 volume was calculated using numerical integration. The H_2 volumes are presented in Table 1. The sol-gel derived $NiFe_2O_4$ generated 100.04, 22.35, 13.65 and 9.63 mL of H_2/g of ferrite loaded in the packed bed reactor during the 1st, 2nd, 3rd and 4th water-splitting step performed at 900°C. Contrasting to this, 124.32, 55.03, 50.36 and 30.69 mL of H_2/g of ferrite was observed during the 1st, 2nd, 3rd and 4th water-splitting step using $NiFe_2O_4/ZrO_2$ powdered material at similar operating conditions. On an average the $NiFe_2O_4$and $NiFe_2O_4/ZrO_2$ powdered material generated 36.42 ml/g/cycle and 64.6 ml/g/cycle, respectively. The H_2 volume generated with $NiFe_2O_4/ZrO_2$ powdered material prepared using ultrasonic mixing is higher as compared with the same material prepared using vortex mixing (47.69 mL/g·cycle) [13].

Table 1 Comparison of H_2 volume generated during four consecutive thermochemical cycles using Ni-ferrite and Ni-ferrite mixed with ZrO_2.

Ferrite	1st	2nd	3rd	4th	Avg H_2 (ml/g)
Ni-ferrite	**100.04**	**22.35**	**13.65**	**9.63**	**36.42**
Ni-ferrite/ZrO_2	124.32	53.03	50.36	30.69	64.6

Figure 5 SEM image of reacted $NiFe_2O_4/ZrO_2$ powdered material.

The SEM image of the reacted $NiFe_2O_4$ that underwent four consecutive thermochemical cycles indicated decrease in specific surface area, increase in the grain size to 2 – 5 μm and decrease in porosity. The SEM image of reacted $NiFe_2O_4/ZrO_2$ powdered material presented in Figure 5 does not indicate significant grain growth and the grain size appears to be much smaller as compared with reacted $NiFe_2O_4$.

Conclusions

Using chloride precursors of Ni and Fe, $NiFe_2O_4$ with nominally phase purity was successfully synthesized using sol-gel technique. Further, $NiFe_2O_4/ZrO_2$ powdered mixture was prepared by adding of ZrO_2 nanoparticles (25 wt% based on $NiFe_2O_4$) to the sol-gel derived $NiFe_2O_4$ powder. The SEM and TEM analysis of both the powders indicated nanoparticle morphology. Both $NiFe_2O_4$ and $NiFe_2O_4/ZrO_2$ powdered materials were used for the H_2 generation in four consecutive thermochemical cycles where water-splitting was performed at 900°C and the regeneration was carried at 1100°C. It was observed that $NiFe_2O_4/ZrO_2$ powdered material prepared using ultrasonic mixing generated higher volume of H_2 (avg. 64.6 ml/g/cycle) as compared with the $NiFe_2O_4$ (avg. 36.42 ml/g/cycle). The grain growth was found to be inhibited in presence of ZrO_2 that resulted in generation of higher H_2 volume.

Acknowledgements

The authors gratefully acknowledge the financial support by the National Science Foundation, Grant No. CBET – 1134570 and the Chemical and Biological Engineering Department at the South Dakota School of Mines and Technology.

References

1. Agrafiotis C., Pagkoura C., Lorentzou S., Kostoglou M., Konstandopoulos A.(2007), *Catalysis Today* **127**, 265.
2. Bhosale R, Shende R, Puszynski J.(2011), *International Journal of Hydrogen Energy* **37**, 2924.
3. Roeb M, Sattler C. Kluser, Monnerie N, Oliveira L, Konstandopoulos A, Agrafiotis C, Zaspalis V, Nalbandian L, Steele A, Stobbe P. (2006), *Journal of Solar Energy Engineering* **128**, 125.
4. Steinfeld A. (2005), *Solar Energy* **78**, 603.
5. Abanades S, Flamant G. (2006), *Solar Energy* **80**, 1611.
6. Fresno F, Yoshida T, Gokon N, Fernandez-Saavedra R, Kodama T.(2010), *International Journal of Hydrogen Energy* **35**, 8503.
7. Bhosale R, Shende R, Puszynski J. (2010), *Journal of Energy and Power Engineering* **4**, 27.
8. Bhosale R, Shende R, Puszynski J. (2010), *International Review of Chemical Engineering* **2**, 852.
9. Yang. H, Zhang X, Ao W, Qiu W. (2004), *Materials Research Bulletin*, **39**, 833.
10. Shi J.L, Deguchi Y, Sakabe Y. (2005), *Journal of Materials Science*, **40**, 5711.
11. Kodama T, Gokon N, Yamamoto R. (2008), *Solar Energy* **82**, 73 (2008).
12. Gokon N, Murayam H, Nagasaki A, Kodama T. (2009), *Solar Energy* **83**, 527.
13. Bhosale R, Shende R, Puszynski J. (2012), Mater. Res. Soc. Symp. Proc. Vol. 1387.

(400d) Synthesis of Ferrite/ZrO2/YSZ Core-Shell Nanoparticles for H2 Generation From Thermochemical Water-Splitting Reaction

Sowmya Yelakanti and Rajesh V. Shende, Department of Chemical and Biological Engineering, South Dakota School of Mines and Technology, Rapid City, SD

H_2 is a green fuel with maximum energy density as compared with the fossil fuels. Thermochemical water-splitting reaction was performed using sol-gel derived Ni-ferrite materials at water-splitting and regeneration temperatures of 700°-1100^oC. After water-splitting reaction, the ferrite material was recovered and characterized using x-ray diffraction, scanning electron microscopy (SEM) and BET surface area analyzer. Although no change in phase composition was observed, a significant increase in grain size and decrease in specific surface area was noticed. When ferrite materials were subjected to thermochemical water-splitting, significant sintering and grain growth was observed that led to decrease in H_2 volume generation during multiple thermochemical cycling operation. To address this issue, thermal stabilization of ferrite material was investigated by performing the synthesis of core-shell nanoparticles where sol-gel derived Ni-ferrite formed a core within ZrO_2/YSZ (Yttria-stabilized Zirconia) shell. The synthesis of ferrite/ZrO_2/YSZ was accomplished using micro-emulsion technique involving the use of two surfactants. In addition, Ni-ferrite/ZrO_2/YSZ powdered material was also prepared using a vortex mixer. These materials were analyzed using SEM, XRD and BET surface area analyzer. The H_2 generation ability of these ferrite nanomaterials was investigated by performing five consecutive thermochemical cycles in the Inconel packed-bed reactor where water-splitting and regeneration were carried out at 700-900^oC and 1100^oC, respectively. The use of ZrO_2/YSZ as grain growth inhibitor or as a shell material in core-shell nanoparticles was affected the H_2 volume generation. The results obtained on grain growth mitigation, core-shell nanostructures and H_2 volume generation from thermochemical water-splitting reaction will be presented in detail.

(400e) Dynamic Model of a Solar Thermochemical Water-Splitting Reactor with Integrated Energy Collection and Storage

Rong Xu and Theodore Wiesner, Chemical Engineering, Texas Tech University, Lubbock, TX

Abstract

Global climate change is the preeminent challenge of the 21st century. Carbon dioxide from fossil fuel combustion is altering the world's environments, potentially in undesirable ways. It is critical to develop carbon-free and environmental friendly energy sources. Molecular hydrogen is currently used as an energy storage medium, and could possibly be used in the future as a transportation fuel.

Water splitting solar thermochemical cycles (WSTCs) are used to decompose water to obtain hydrogen through redox reactions. Of the approximately 200 WSTCs identified thus far, the Cu-Cl cycle (**Figure 1**) is particularly attractive, since the maximum temperature required of any step within the cycle (530 ^{o}C) is much lower than some of the comparable WSTCs such as the Ni-Fe cycle (1100 ^{o}C) or the Zn-ZnO cycle (800 ^{o}C). The Cu-Cl cycle requires a five-step reaction cycle to generate hydrogen. It is the fifth step, which is the decomposition of copperoxychloride ($CuOCuCl_2$) into cuprous chloride (CuCl) and oxygen, that requires the maximum temperature (530 ^{o}C), and it is this step for which we created a solar reactor design.

The traditional solar reactor design has not compensated for solar variation throughout day and night in a way to achieve continuous operation. Instead, for these reactors, the operation routinely starts up and shuts down according to sunshine duration (**Figure 2**).

We propose a new solar reactor system, shown in **Figure 3**. Decomposition of CuOCuCl2 occurs within the central reactor tube, from which, the mixed products flow into a separator for further reaction. The inner annular shell around the reaction pipe contains the heat transfer medium (HTM), which flows in a countercurrent direction relative to the reactor flow. A spectrum-selective coating is painted on the surface of this shell. The high absorption and low emissivity coating absorbs the concentrated solar radiation. The outermost cylinder is a vacuum tube constructed from glass, which is used to avoid heat convection losses to the atmosphere. Concentrated solar power is provided to the reaction system by a reflective parabolic trough (not shown). The HTM is recycled through a tank which functions as a thermal storage device. The tank buffers the variation of the temperature in HTM and reactor, saves energy during periods of solar insolation and releases heat during times of low solar insolation. The collection of excess heat during insolation ensures a steady hydrogen production during periods of low insolation.

The design described above was modeled by a system of partial differential equations, in which the reactor temperature, temperature of the HTM, and conversion of the reactant ($CuOCuCl_2$) varied with time and axial position within the reactor. The model is a one-dimensional plug flow reactor (PFR) model. The PDEs were discretized into ODEs which were then solved at a series of design parameters and operating conditions within MathCAD software. The goals to be achieved through the choice of the design and operating parameters were the following: (i) to buffer the effects of transient solar insolation in order to maintain a steady operation at 100% conversion, (ii) to minimize the fluctuation of the conversion once the steady operation was achieved. The main constraint to be respected was the maximum operation temperature of the molten salt (700 ^{o}C).

First, we compared the directly irradiated reactor (Base case, **Figure 4**) with the jacketed system, and found that the

HTM did attenuate the effect of time varying solar irradiation upon the conversion of the decomposition reaction. Then, we modified the design parameters, such as the solar concentration ratio as well as operating parameters, such as the mass flow rate of HTM, though which we achieved a simulated 100% conversion(Optimal case, **Figure 5**). Less than eight hours after reactor start-up, steady operation was achieved, and this continued for the length of the simulation (three 24-hour days). At this steady operation point, we modeled a hydrogen production of 3.5 kg/day. For simplification of this preliminary study, the molten salt flow rate was kept at a constant value, although in our future work, the effects of using a higher solar concentration ratio along with an optimized transient flow rate of the heat transfer medium will be studied to see if the footprint of the HTM storage tanks could be reduced.

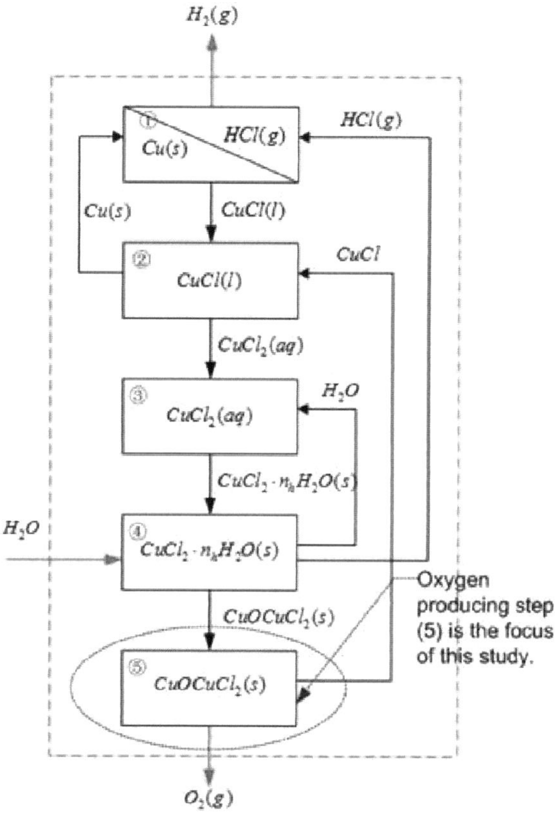

Figure 1 Schematic of the 5-step CuCl2 WSTC

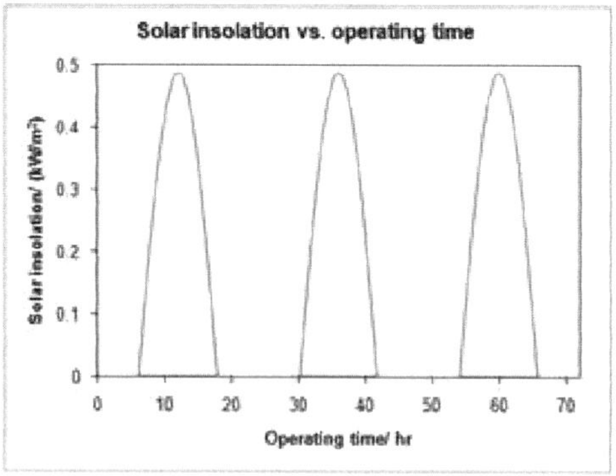

Figure 2 Insolation vs. operation time

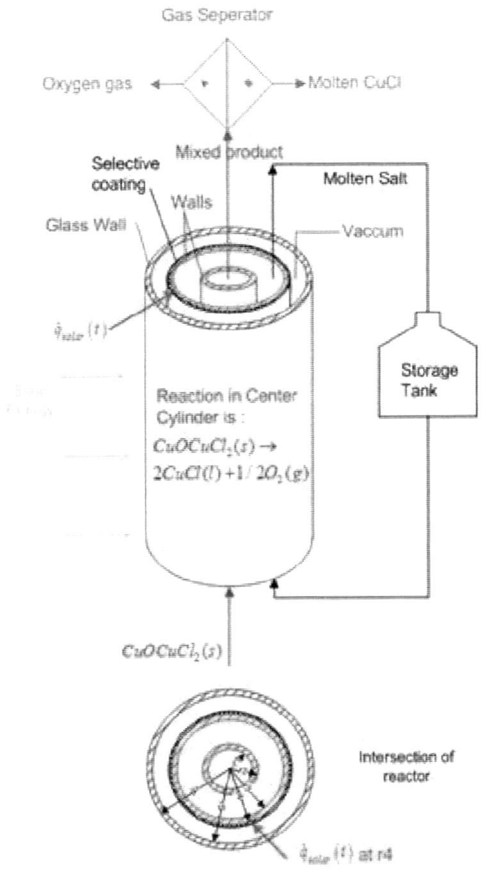

Figure 3 Schematic of reactor and energy storage system

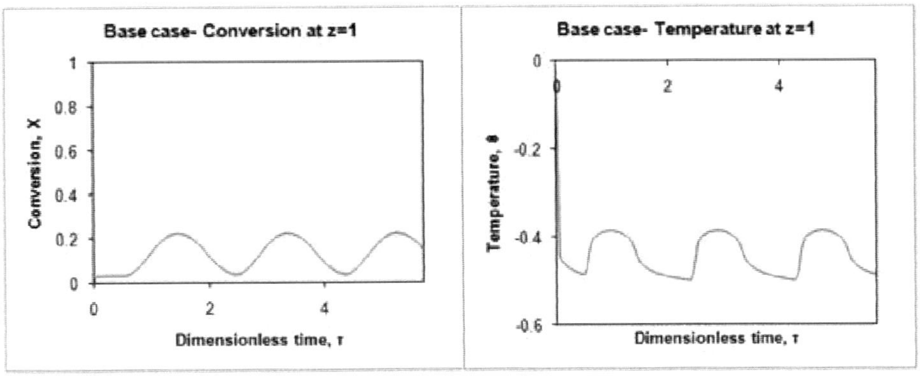

Figure 4 Conversion and temperature at reactor outlet in base case

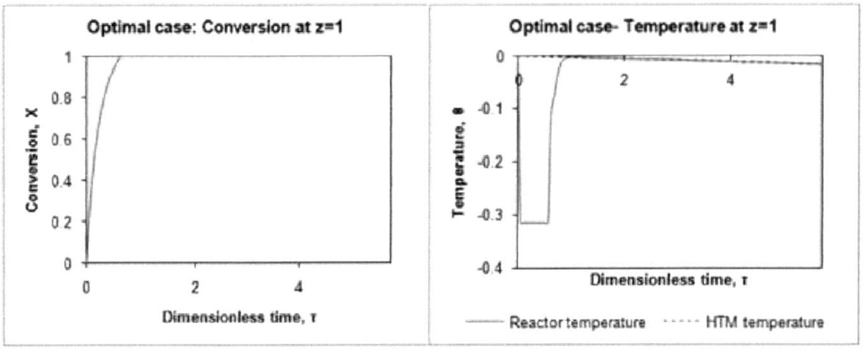

Figure 5 Conversion and temperature at reactor outlet in the optimal case

(486a) Development of Metal-Oxide-Semiconductor Photoelectrodes for Photoelectrochemical Water Splitting

Daniel V. Esposito, Alec Talin and Thomas Moffat, National Institute of Standards and Technology, Gaithersburg, MD

Photoelectrochemical (PEC) water splitting is a promising pathway for renewable H_2 production, but there remains a great need for improving the efficiency and stability of semiconducting photoelectrodes. One promising approach to achieving high efficiency and good electrochemical stability is the metal-oxide-semiconductor (MOS) photoelectrode design.[1] This MOS arrangement generally consists of catalytic metal structures, or collectors, deposited on an oxide-covered semiconductor absorber layer. The critical element of this design is the thin oxide layer, which protects the semiconductor from the potentially corrosive electrolyte while mediating minority carrier tunneling between the semiconductor and collectors. The oxide layer must be ultra-thin (1-3 nm) and exhibit minimal defects at the SiO_2/Si interface in order to facilitate efficient tunneling and maximize photovoltage. Si is a commonly used photovoltaic material that is also attractive for use in MOS photoelectrodes, but Si-based MOS electrodes demonstrated to date have greatly suffered from poor oxide quality, resulting in interfacial trap states that drastically lower MOS photovoltage and overall water splitting efficiency.[2]

In order to enhance the solar-to-hydrogen conversion efficiencies (η_{STH}) of Si-based MOS photoelectrodes, our research efforts have focused on two areas. First, well-defined and ordered metallic structures have been synthesized on high quality SiO_2 tunneling layers, grown by rapid thermal annealing (RTA) on p-Si(100). Thermally-grown SiO_2 is known to exhibit low interfacial defect densities,[3] meaning that voltage losses incurred by carrier tunneling via the associated mid-gap trap states can be minimized.[2] In this work, we show that the use of thermally grown SiO_2 layers for Pt-SiO_2-Si photoelectrodes produces a substantial increase in photocurrent onset potential. The well-defined and ordered nature of the metal components in these photoelectrodes provides the additional benefit of being able to optimize the spacing and coverage of collector structures. By systematically varying these two parameters, we show how optical and kinetic overpotential losses can be balanced to maximize η_{STH} values for a given substrate/oxide combination. The second thrust of this work is using in-situ photoelectrochemical measurements with high spatial resolution to aid in optimization of MOS structures and gain a better understanding of charge transfer phenomena in those structures. This talk will specifically discuss the use of photoelectrochemical microscopy (PECM) for micro-scale mapping of external quantum efficiency and determination of important device parameters such as the minority carrier diffusion length. PECM measurements indicate substantial room for improvement in macro-scale photoelectrode performance and demonstrate the powerful utility of PECM in the design and analysis of MOS photoelectrodes.

References

[1.] H.J. Lewerenz, et al., *Electrochem. Acta*, **2011**, 56, 10726.

[2.] A.G. Munoz and H.J. Lewerenz, *ChemPhysChem*, **2010**, 11, 1603.

[3.] R.R. Razouk and B.E. Deal, *J. Electrochem. Soc.*, **1979**, 126, 1573.

(486b) Novel Ternary and Doped Transition Metal Oxide Oxygen Evolution Electro-Catalysts for Hydrogen Production Via PEM Based Water Electrolysis

Karan Kadakia[1], Moni Datta[2], Prashanth Jampani Hanumantha[1] and Prashant Kumta[1,2,3], (1)Chemical and Petroleum Engineering, University of Pittsburgh, Pittsburgh, PA, (2)Bioengineering, University of Pittsburgh, Pittsburgh, PA, (3)Mechanical Engineering and Materials Science, University of Pittsburgh, Pittsburgh, PA

Hallmark of a sustainable world is one wherein energy demands are consistently met with renewable energy sources. In this regards, hydrogen is often touted as the "magic" fuel of the future if it can be derived from renewable systems such as solar energy as the source of electrical potential required for splitting water by electrolysis. Hydrogen one of the most abundant and lightweight element in the universe is also a clean and efficient energy carrier in contrast to the fossil fuels which are continuously depleting by the day. Hydrogen is considered the "forever fuel" since it can be produced from any primary energy fuel. A hydrogen based energy system can thus be considered benign and very much desirable for delivering high quality energy services in an efficient, clean and safe manner contributing to achieving the sustainability goals. Hydrogen is mainly produced currently from fossil fuels via steam reforming or partial oxidation of natural gas, which generates large amounts of carbon dioxide in turn contributing to global warming and ozone depletion. Production of hydrogen by electrolysis of water is thus considered a potential, practical solution. Electrolysis is less efficient than the chemical pathway but offers virtually no pollution or toxic byproducts if the electric current is generated using renewable energy sources. Currently established technologies for producing hydrogen require significant improvements in their technical and economic performance (efficiency and costs) if hydrogen is to be produced for energy use. High capital costs are thus incurred due to use of expensive materials, relatively small systems, reasonably low efficiencies, customized power electronics, coupled with labor intensive fabrication.

For proton exchange membrane (PEM) electrolysis cells, use of non-precious metal catalysts for electrodes with high activity and durability would significantly decrease the overall capital costs. While intense research is being conducted for identifying a completely non-noble metal catalyst for PEM system, approaches to decrease noble metal loading is extremely critical. Decreasing the noble metal loading as well as improving the catalytic activity and durability could be achieved using novel synthesis techniques to generate high surface area nano electro-catalyst, or identifying approaches to generate low cost support structures (diluents) for electro-catalyst. The present work is thus carried out to identify a novel support for electro-catalyst which can appreciably decrease the precious metal loading without compromising the electrochemical activity and correspondingly improving the corrosion stability of noble metal oxide electro-catalyst.

Non-noble metal oxides though chemically stable are known to display however, no catalytic activity for oxygen reduction processes. Mixed oxides obtained by the addition of economically cheaper alternatives to noble metal oxides could offer significant cost effective options if the systems exhibited catalytic activity, chemical stability, and electronic conductivity similar to the currently accepted gold standard, IrO_2 electro-catalysts used in PEM based water electrolysis. On the grounds of the very promising performances shown by both IrO_2-SnO_2 and IrO_2-Nb_2O_5 mixtures as an anode electro-catalyst, the present investigation has been conducted to explore the synthesis and characterization of novel composite-ternary transition metal oxide systems. Also, in the present research pure and group VII doped SnO_2 solid solution with IrO_2 has been explored as electro-catalysts. Results to date indicate that compositions containing 80% reduction in noble metal oxide exhibit comparable electrochemical activity to pure IrO_2.

(486c) New Reactor Concept for Photocatalytic Hydrogen Evolution Under Sunlight Irradiation

Marc Schroeder, Institute of Chemistry, TU Berlin, Berlin, Germany, Michael Schwarze, Fachgruppe Technische Chemie, Technische Universitaet Berlin, Berlin, Germany, Kamalakannan Kailasam, TU Berlin, Berlin, Germany, Prof. Dr. Arne Thomas, TU Berlin, berlin, Germany and Prof. Dr. Reinhard Schomäcker, Institute of Chemistry, Secr. TC8, Technical University of Berlin, Berlin, Germany

New reactor concept for photocatalytic hydrogen evolution under sunlight irradiation

M. Schröder, M. Schwarze, K. Kailasam, A. Thomas, R. Schomäcker

Technische Universität Berlin, Institut für Chemie, Straße des 17. Juni 124, 10623 Berlin

The depletion of fossil fuels and increase of global warming lead to a high demand for alternative energy sources that are clean, cheap and renewable. In addition to wind, water and solar energy, electricity generation via fuel cells is one of the most promising "green energy" sources of the future and therefore hydrogen will become more and more important. But the production of hydrogen by petrochemical routes which are the most applied technologies, are based on fossil fuels and emit CO_2. Thus, the production of hydrogen from water via sunlight is considered as a "Dream Reaction" for hydrogen evolution. Currently different groups are working with this discipline of "green energy" to develop new photocatalysts which are able to split water into hydrogen and oxygen or to reduce water to hydrogen with the assistance of sacrificial reagents. Within the interdisciplinary project "Light2Hydrogen", which is funded by the "Bundesminesterium für Bildung und Forschung", homogeneous and heterogeneous photocatalysts are synthesized and analyzed for efficient utilization of solar energy in photocatalytic hydrogen production. To date, homogeneous catalyst in this project have reached a hydrogen production rate of up to 20 mL / h [1] and the most efficient heterogeneous photocatalyst, which is a mesoporous carbon nitride (mpg-C_3N_4), reaches a hydrogen production rate of up to 2 mL/h (Xenon lamp) [2]. To generate power from the produced hydrogen, for example in combination with a fuel cell, it is important to design a scalable photoreactor which is combinable with other devices like hydrogen storage and electricity generation. Therefore, a new photoreactor was designed which (a) has a planar window for a defined irradiation with less reflection, (b) uses the pressure increase to follow the reaction and (c) allows a controlled release of the produced hydrogen during the process. The reactor allows quantitative testing of different catalyst materials, kinetic investigations and to calculate the degree of efficiency based on the irradiated energy.

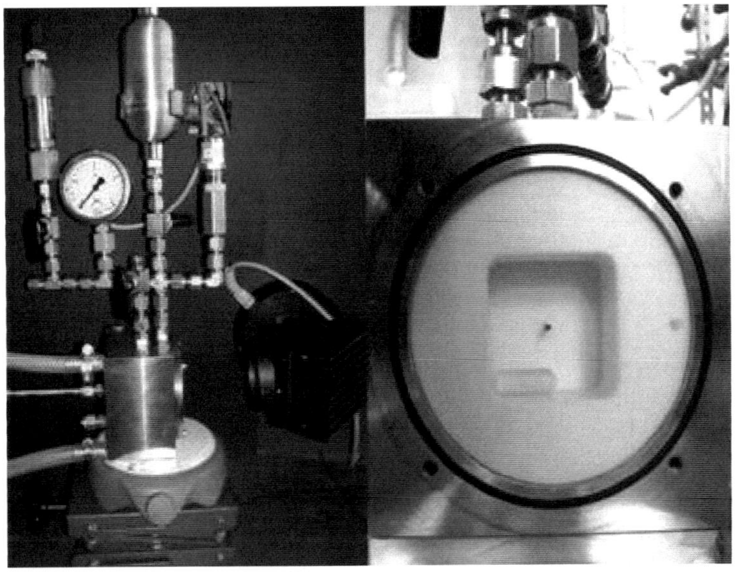

Figure 1: Setup for photorcatalytic investigations (left) and front view of reactor (right).

In this new photoreactor, kinetic investigations of mesoporous carbon nitrides as photocatalysts were performed under sunlight simulation. A hydrogen evolution rate of 0,55 mL/h is obtained under optimized reaction conditions. The catalyst is stable and photocatalytic activity is retained after several dark periods as shown in figure 2.

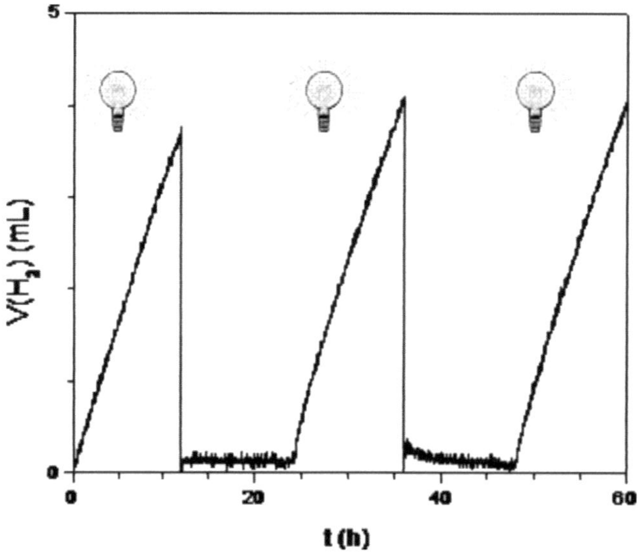

Figure 2: Volume of hydrogen as function of time for several light and dark periods

This contribution will introduce the new photoreactor concept and show some results of the kinetic investigations.

[1] F. Gärtner et al., Chem. Eur. J. 17, 2011, 6998-7006

[2] K. Kailasam et al., Energy Environ Sci 4, 2011, 4668-4674

(486d) Biological Templates for Anti-Reflectance Surface and Current Collectors for Photoelectrochemical Cell Applications

Chia-Ying Chiang, Chemical and Biomolecular Engineering, University of Maryland, College Park, MD, Jillian Epstein, Department of Chemical and Biomolecular Engineering, University of Maryland, college park, MD, Adam Brown, Institute for Bioscience and Biotechnology Research, College park, MD, James N. Culver, University of Maryland Biotechnology Institute, College Park, MD and **Sheryl H. Ehrman**, Department of Chemical and Biomolecular Engineering, University of Maryland, College Park, MD

Three dimensional (3D) structures such as nanowires, nanotubes, and nanorods have the potential to increase surface area, reduce light reflection and shorten charge carrier transport distances. Assembly of such structures thus holds great promise for enhancing photoelectrochemical solar cell applications. In this study, genetically modified *Tobacco mosaic virus* (TMV1cys) was used to form self-assembling 3D nanorod current collectors and low light reflecting surfaces. Photoactive CuO was subsequently deposited by sputtering onto these patterned nanostructures and examined for photocurrent activity. An assembled surface structure with period of ~500 nm, i.e. TMV1cys concentration at 10^{-3} mg/mL, was demonstrated to suppress light reflection from 13% down to 3% leading to a higher efficiency in the utilization of incident photons. CuO thicknesses of 520 nm on TMV1cys patterned current collectors produced the highest photocurrent density of 3.15 mA/cm^2 yet reported for a similar sized CuO system. Thus the combined effects of reducing charge carrier transport distance, increasing surface area and the suppression of light reflection makes these virus assembled surfaces ideal for photoelectrochemical applications.

UV AND VISIBLE LIGHT PHOTOCATALYTIC DEGRADATION OF GLUCOSE BY ZrO_2 DOPED TIO_2/ZnO NANOMATERIAL

Anu Shende, Evan Sellers, Rajneesh Jaswal and Rajesh V Shende
South Dakota School of Mines and Technology,
Rapid City, SD, USA

Abstract

UV and visible excitable ZrO_2 doped TiO_2/ZnO photocatalyst was prepared by sol-gel assisted surface templating approach. TiO_2/ZnO gels were prepared with 'Zr' in 0.5 wt % and 10 wt% and calcined at 300 °C and 1000 °C. As-calcined materials were characterized by BET surface area analyzer, transmission electron microscopy (TEM) and UV-Visible diffuse reflectance spectroscopy. The calcined materials were coated with 1% Pt using the chloroplatinic acid and their effectiveness on degradation and hydrogen generation from water-splitting and glucose oxidation was studied under simulated solar irradiation.

Introduction

TiO_2 is the most widely studied photocatalyst with a potential for commercial application in biomass reforming if only researchers find a way to overcome its two major limitations namely a) lack of photosensitivity to the wavelengths in the visible range of the solar spectrum and b) short exciton diffusion length, allowing only those carriers generated within space charge layer contribute to the photocurrent [1,2]. One of the ways to reduce the band gap and increase photosensitivity in the visible range ($\lambda > 400$ nm) is to dope TiO_2 with metal oxides. TiO_2 nanomaterials doped with metal oxides have shown increased photocatalytic efficiencies [3-5]. Of particular interest is the doping of TiO_2 with the zirconium oxide and zinc oxide [6]. Zirconium reportedly contributes to an increase in surface area of the TiO_2 nanoparticles and also, partially prevent the e^-/h^+ recombination [7]. Zirconium also increases the thermal stability of the photocatalytically superior anatase phase of the TiO_2. The semiconductor ZnO has a wide energy band gap of 3.4 eV compared with the 3.2 eV of anatase TiO_2 [8]. However, the UV–vis absorption of the ZnO/TiO_2 showed a red-shift probably because of replacement of the oxygen in the TiO_2 lattice and surface with the Zn^{2+} ion thus creating defects and an energy level in the TiO_2 band gap [9]. This energy band gap can be used to excite electrons to the conduction band to inhibit the recombination of electrons and holes.

Oxidation of organic compounds (e.g. glucose) takes place in the presence of oxygen (air) with the participation of photo-generated holes, which act either directly or indirectly via generation of hydroxyl radicals (OH^-), and ultimately lead to the production of H_2, H_2O and CO_2. In this regards, the use of TiO_2 coupled with ZnO and further doped with Zr will be very useful as in such a material, the band gap offset at different interfaces and additional induction of defects due to Zr can lead to enhanced charge separation and prolonged lifetimes that may enhance the photo induced reaction on catalyst surfaces.

In this investigation, we sought to understand the contribution of Zr in stabilizing TiO_2 at various calcination temperature and in improving photocatalytic efficiency of ZnO/TiO_2. Using the sol-gel method, we have doped ZnO/TiO_2 with 0.5 wt% and 10 wt% of zirconium and studied its role in improving the photocatalytic efficiency of ZnO/TiO_2 calcined at different temperatures.

The UV-visible light driven photocatalytic conversion of glucose to hydrogen was used to evaluate the performance of the doped materials.

Experimental

Preparation of ZnO/ZrO_2 doped TiO_2

To prepare ZnO/ZrO_2 doped TiO_2, 5 gms of Titanium isopropoxide was mixed with 10 ml ethanol. The mixture was acidified with 100 ul of 12 N sulphuric acid. In a second beaker 0.5 gms of $ZnCl_2$ was mixed with various amounts of Zirconiumn-n propoxide to final weight percent of Zr ranging from 0.5 to 10% . In another beaker, Pluronic 123 was dissolved in 10 ml ethanol. The contents of the three beakers were mixed. Ammonium hydroxide (2ml) was added to the mixture and it was allowed to gel overnight. The material was dried in a hot air oven and calcined in a muffle furnace at various temperatures. The photocatalysts were coated with 1% platinum using hexachloroplatinic acid.

Characterization of Photocatalysts

The photocatalyst material were characterized by TEM and BET surface area analysis. Powdered photocatalyst was directly mounted on stainless steel stubs coated with an adhesive carbon tape, and observed under JEOL JSM-840A scanning electron microscope equipped with an Oxford Instruments energy-dispersive X-ray (EDX) analyzer for elemental analysis. For TEM analysis the sample was evenly dispersed in isopropyl alcohol by sonication for 60 minutes after which it was mounted on copper grid with carbon support film and dried in air at room temperature. TEM images were obtained at 300 kV using Hitachi H-7000 FA TEM. To obtain X-ray diffraction pattern, the sample was scanned from 5° to 90° with a scanning speed of 1° per minute by Rigaku Ultima-Plus X-ray diffractometer.

Photocatalysis of Glucose

In a photocatalysis reactor shown in the Figure 1, 40 mg of glucose was added to 4 mg of photocatalyst dispersed by sonication in 10 ml water. A Teflon coated magnetic stir bar was inserted. The media was purged with nitrogen for 15 min. The photoreactor was exposed to a solar simulator fitted with air mass filter 1.5 G. The power density measured at the center of the reactor placed on a magnetic stirrer was $100W/cm^2$. The gas phase analysis was made by injecting a 100 ul sample withdrawn from the headspace into a gas chromatograph.

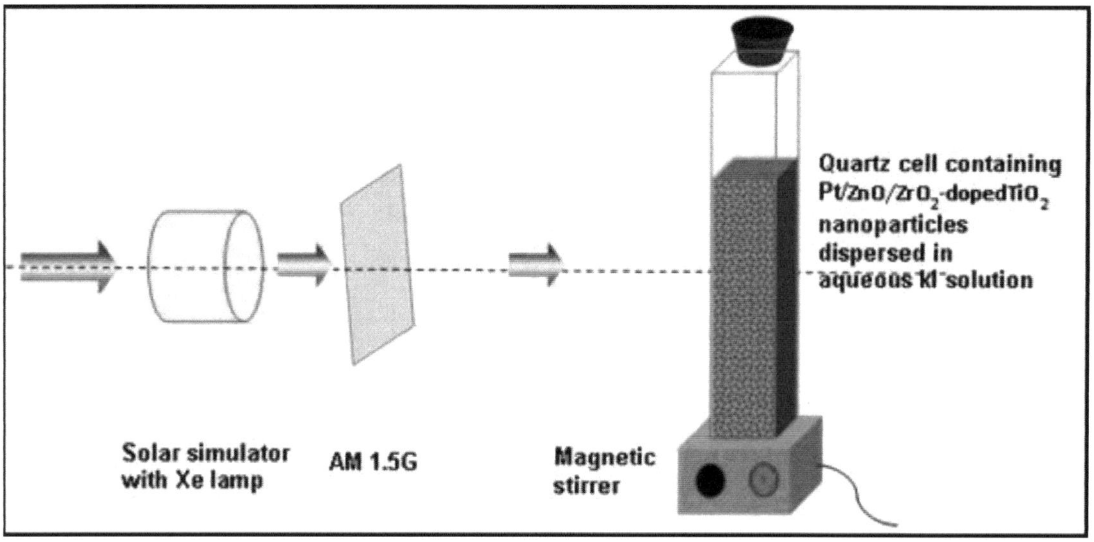

Figure 1 Experimental photoreactor set-up for water-splitting and photocatalysis of glucose.

Results and Discussion

Table 1 shows BET surface area of ZnO/TiO_2 prepared with 0.5 wt% and 10 wt% Zr calcined at two different temperatures. The gels prepared with Pluronic 123 templating were rapidly calcined at a ramp rate of 40°C/min with no hold conditions at the calcination temperature. It can be observed that for the materials calcined at 300°C, the BET surface area is 36-41 m^2/g, which decreased significantly to about 2 m^2/g when calcination temperature was increased to 1000°C. The data indicates increase in particle size with calcinations temperature. This observation is consistent with the published literature.

Table 1 BET specific surface area of ZrO_2 doped ZnO/TiO_2 materials prepared with different "Zr' loadings

Zirconium wt%	Calcination temp (°C)	Specific Surface area, m^2/g
10	1000	2.27
10	300	36.68
0.5	1000	2.16
0.5	300	41.13

A TEM image of ZrO_2 doped ZnO/TiO_2 prepared with 0.5 wt% Zr is shown in Figure 2, which shows spherical nanoparticles ranging in size from 7-10 nm.

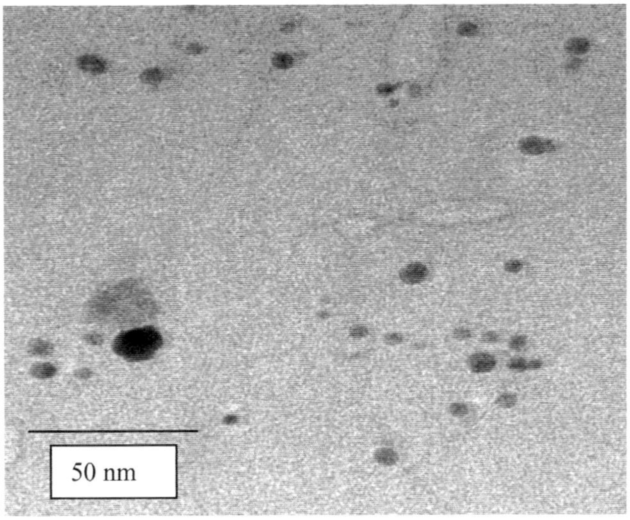

Figure 2 TEM image of ZrO_2 doped ZnO/TiO_2 prepared with 0.5 wt% Zr and calcined at 300°C.

Diffuse reflectance spectroscope equipped with an integrating sphere was used to record the diffuse reflectance spectra (DRS) of photocatalytic materials in 200- 500 nm and $BaSO_4$ was used as a reference. The diffuse reflectance spectra of ZrO_2 doped ZnO/TiO_2 prepared with Zr at 0.5 wt% and 10 wt% and calcined at different temperatures are shown in Figure 3. For comparison, diffuse reflectance spectra of nominally phase pure TiO_2, ZnO and ZrO_2 are included.

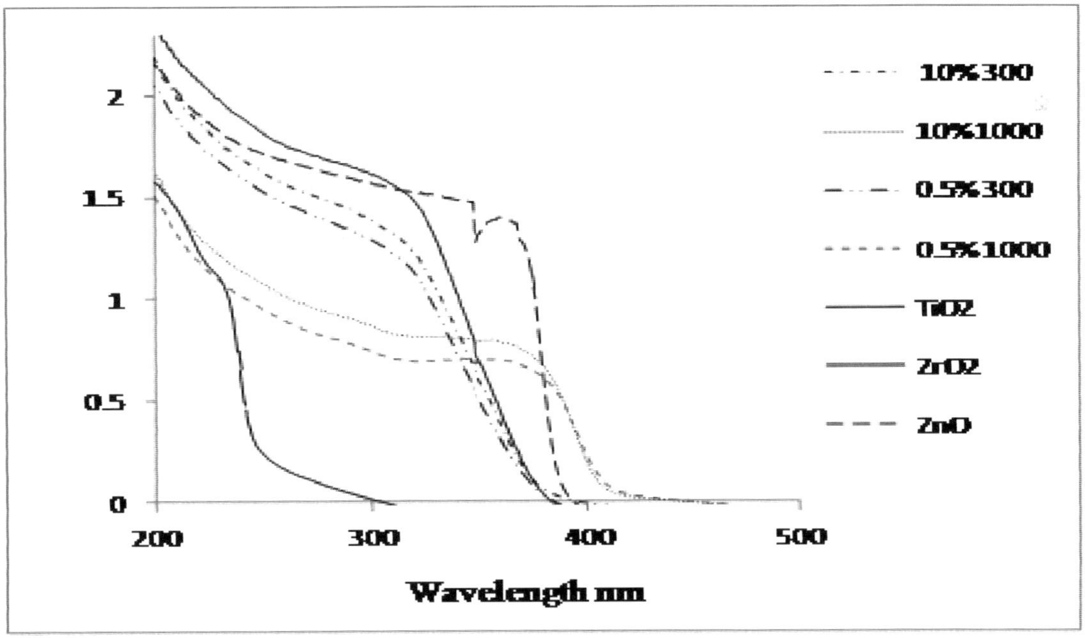

Figure 3 Diffuse reflectance spectra of ZnO, ZrO_2, TiO_2, and ZrO_2 doped ZnO/TiO_2 prepared with 0.5 wt% and 10 wt% Zr and calcined at 300°C and 1000°C.

It can be observed that as expected, materials doped with either 0.5 wt% or 10 wt% Zr and calcined at 1000°C showed absorption shift in visible range.. However, the materials calcined at 300°C did not show any shift in visible range. This suggests that for solid solution to form, higher temperatures are desired. It should be noted that the specific surface area of the materials calcined at 1000°C was about 2 m^2/g indicating increase in particle size. The data clearly demonstrates challenges to achieve desired phase formation without sufficient energy input crossing thermal activation barrier.

The photocatalytic effectiveness of ZrO_2 doped ZnO/TiO_2 materials towards hydrogen generation from aqueous glucose solution were investigated and the results obtained are shown in Figure 4. The material prepared with 0.5 wt% Zr and calcined at 300°C showed highest relative hydrogen yeild of about 52 vol% whereas when the material prepared with 10 wt% Zr and calcination temperature of 300°C, showed reduced photocatalytic efficiency producing only 20% relative hydrogen yield. This shows that there exists an optimum Zr doping in ZnO/TiO_2 material over which the excitons recombine. For the materials prepared with 0.5 vol% or 10 wt% Zr and calcined at 1000°C, negligible amount of hydrogen was observed. As the specific surface area of these materials is about 20 times lower than those calcined at 300°C (Table 1), it appears that the lower yields of hydrogen was limited by the number of catalytic sites available.

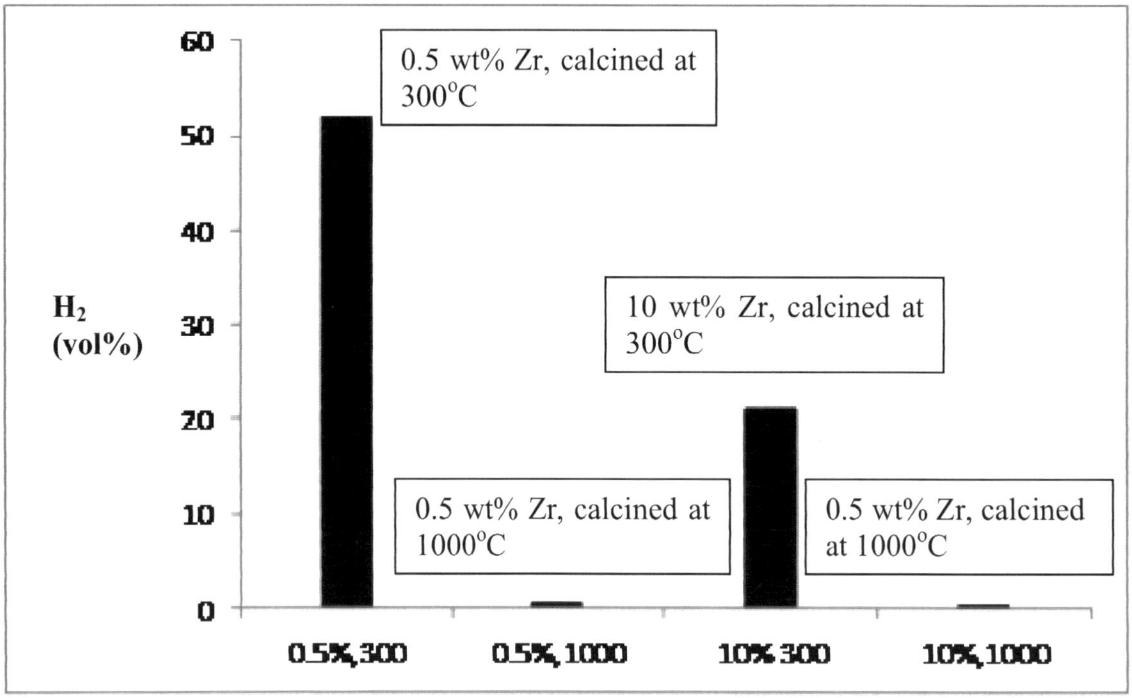

Figure 4 Comparison of hydrogen vol% observed during photocatalytsis of aqueous glucose solution using ZrO_2 doped ZnO/TiO_2 prepared with 0.5 wt% and 10 wt% Zr and calcined at 300°C and 1000°C.

Conclusions

ZrO_2 doped ZnO/TiO_2 materials prepared with Zr loadings 0.5 wt% and 10 wt% calcined at 300°C indicated specific surface area (SSA) of 36-41 m2/g, whereas for these materials calcined at 1000°C, the SSA reduced significantly to about 2 m^2/g. TEM analysis of of ZrO_2 doped ZnO/TiO_2 prepared with 0.5 wt% Zr and calcined at 300°C showed particles in the size range of 7-10 nm. The DRS study revealed visible light absorption threshold for the materials calcined at 1000°C. These materials when evaluated for hydrogen generation from aqueous glucose solution showed negligible hydrogen. One of the reasons for this is significant reduction in surface area and therefore the number of active sites that limited the effectiveness of photoinduced reactions. Contrary to this, the materials calcined at 300°C showed remarkable hydrogen yeilds from 20-52 vol% for 10 wt% and 0.5 wt% Zr doping, respectively. This suggest that there is optimum Zr doping above which excitons can recombine and limit the surface reaction.

Acknowledgement

Authors gratefully acknowledge the financial support by the Chemical and Biological Engineering at SDSM&T.

References

1. J.H. Park, S. Kim, A.J. Bard, Novel carbon-doped TiO_2 nanotube arrays with high aspect rations for efficient solar water splitting, *NanoLetters*, **6;** 24 (2006).
2. S. Sato, J. White, Photodecomposition of water over Pt/TiO_2 catalyst, *Chemical Physics Letters* **72**, 83 (1980).
3. Maolin Zhang , Taicheng An, Xiaolu Liu, Xiaohong Hu, Guoying Sheng, Jiamo Fu Preparation of a high-activity ZnO/TiO_2 photocatalyst via homogeneous hydrolysis method with low temperature crystallization, *Materials Letters*, **64** (2010) 1883.
4. GR Bamwenda, S Tsubota, T Nakamura, M Haruta (1995) Photoassisted hydrogen production from water-ethanol solution- a comparison of activities of Au-TiO_2 and Pt-TiO_2, *Journal of Photochemistry and Photobiology A: Chemistry* **89**: 177.
5. S. Sakulkhaemaruethai, A. Kitiyanan, S. Yoshikawa, Photocatalytic activity of ZrO_2-doped TiO_2 catalysts prepared by a surfactant assisted templating method, *Ceramic Processing Research*, **7** (10 (2006) 10-13.
6. A.R. Shende, R. Jaswal, R.V. Shende, ZnO/ZrO_2 doped TiO_2 mediated photocatalytic pretreatment of lignin for microbial conversion to value added products, Nanomaterials and Energy, 2012 (submitted).
7. K.Y. Jung, S.B. Park, Photoactivity of SiO_2/TiO_2 and ZrO_2/TiO_2 mixed oxides prepared by sol-gel method, *Mater. Lett.* **58** (2004) 2897.
8. R.S. Mane, W.J. Lee, H.M. Pathan, S-H. Han, Nanocrystalline TiO/ZnO thin films: fabrication and application to dye sensitized solar cells, *The Journal of Physical Chemistry B*, 2005, **109** (51) 24254.
9. Y.B. Han, J.B. Han, Z.H. Hao, Band gap shift and the optical nonlinear absorption of sputtered ZnO-TiO2 films, *J. Nanoscience and Nanotechnology*, **11** (6) 2011, 5024.

(605a) Mother Nature Vs. Good Engineering Design

Louisa Nara, AIChE/CCPS, New York, NY

Climate change, natural disasters and external events can affect the integrity of operations and infrastructure despite design for safe and effective operations. As process safety and risk professionals, we must be vigilant to assess changing conditions and be able to adapt to prevent catastrophic consequences.

How do you Implement Process Safety? Take the FUN out of it!

Peter N. Lodal
Technical Fellow
Eastman Chemical Company
P.O. Box 511, Building 18
Kingsport, TN 37662
pnlodal@eastman.com

Prepared for Presentation at

AIChE Annual Meeting

Pittsburgh, PA

October 28-November 1, 2012

UNPUBLISHED

How do you Implement Process Safety? Take the FUN out of it!

Peter N. Lodal
Technical Fellow
Eastman Chemical Company
P.O. Box 511, Building 18
Kingsport, TN 37662
pnlodal@eastman.com

Keywords: Process Safety Implementation

Abstract

This paper provides a slightly different approach to implementing process safety than is conventionally applied. Concepts from areas that have nothing to do with chemical engineering or process safety are applied in a highly unusual fashion.

Introduction

When I first was asked to speak on the topic of implementing process safety at a major US chemical facility last spring at the Global Congress of Process Safety, I struggled with how best to get my arms around such a wide-ranging subject. Then, when I was told the slot I had to speak was immediately after lunch, my goals became much less lofty—how do I keep my audience awake?

In the process, I dug up, and connected, a number of different ideas on how best to implement something as "soft" as process safety, and hit on the theme which I use as the title for my talk. I realize this will involve some explanation. So, let us begin with the Grandfather of Process Safety—Leo Tolstoy. Yes, THAT Leo Tolstoy. Although old Leo died in 1910, just two years after the founding of AIChE, he articulated a concept in his novel *Anna Karenina*, from which we will begin our journey.

Dr. Jared Diamond, in his book "Guns, Germs and Steel; the Fates of Human Societies" (1) articulates the Anna Karenina Principle (AKP), from the first line of Tolstoy's 1878 novel:

"Happy families are all alike; every unhappy family is unhappy in its own way."

Dr. Diamond extends this principle to evolutionary biology and the success or failure of civilizations. He writes:

"By that sentence, Tolstoy meant that, in order to be happy, a marriage must succeed in many different respects; sexual attraction, agreement about money, child discipline, religion, in-laws, and other vital issues.

Failure in any one of those essential respects can doom a marriage even if it has all the other ingredients needed for happiness."

I believe this principle can be extended to understanding lots of other things besides marriage and family; namely, process safety. Let me explain; paraphrasing Tolstoy:

"Successful chemical processes are all alike; every unsuccessful chemical process fails in its own way."

If we believe this statement, then a successful chemical process must succeed at a number of things simultaneously—that is, they all look "alike" in the sense that they all succeed at the same things. This suggests some essential elements of a successful chemical process. Although your list may differ slightly from mine, the following things must all go right for a process to succeed in the long term:

- The process must be thermodynamically possible—no alchemy allowed

- The process must be sustainable, in that it:

  - Meets a market need—someone wants to buy it
  - Meets some minimum financial criterion (you can make it at a sufficient profit to stay in business)
  - Consistent raw materials availability
  - Intellectual property protection

- The process must be legally acceptable

  - Not specifically illegal or criminal (e.g. drug dealing)
  - Meets applicable regulatory requirements

- The process must be socially acceptable

  - Safe--Doesn't hurt people (employees or neighbors)
  - Clean--Doesn't hurt the environment (pollution, smell, etc.)

So, if you believe the AKP as applied to process safety, you should adopt the mindset that safety is just as important as sustainability and legal and social acceptability **—not more, not less—** as all successful chemical processes have to succeed at all of these elements consistently, if not simultaneously. This conclusion will necessitate a cultural change, not only on the part of business managers, but on the part of process safety professionals as well. What kind of change has to happen?

The OECD has recently issued a report entitled: "Corporate Governance for Process Safety: Guidance for Senior Leaders in High Hazard Industries" (2). The report, which is very well done, contains this statement:

"Safe operation and sustainable success in business cannot be separated. Failure to manage process safety can never deliver good performance ***in the long term****, and the consequences of getting control of major hazards wrong are extremely costly." (my emphasis added).*

So, what about the short term? Why, if effective process safety management is so beneficial, do leaders not practice it ***in the short term***?

This is a question that has been debated endlessly in other venues:

- Why do people smoke?
- Why do people use drugs?
- Why don't people wear seat belts?

- Why don't motorcyclists wear helmets?

The answer to all of these questions rests on a behavioral fact that we would rather ignore: There is often short-term benefit for NOT doing what will benefit us in the long term; i.e., Smoking and drug use produce short-term pleasure for the users. In the same vein (pardon the pun), minimizing process safety can produce tangible short term benefits to the non-practitioners;

- Profits
- Bonuses
- Promotions

Thus, we are faced with a huge cultural hurdle to overcome. So, how do we go about changing that culture?

Cultural changes require behavioral changes; therefore, we must apply some behavioral based safety (BBS) principles to process safety. Process safety professionals tend to put BBS into the realm of industrial or personal safety. So, how do we apply a behavioral model appropriately to process safety, without losing focus on the unique aspects of process safety?

Aubrey Daniels (3) is a clinical psychologist who founded his own company (ADI), with the mission of applying scientifically-proven laws of human behavior to optimize performance. Daniels says:

"People are the most important element of every organization. With the complex challenges facing American business leaders every day, the science of human behavior cannot be ignored. Rather, it must be the starting place for every decision we make, every new technology we apply, and every initiative we employ."

The heart of Daniels' approach is what he terms the ABC Model

- **A**ntecedents
- **B**ehaviors
- **C**onsequences

Antecedents are what create favorable circumstances for a particular behavior to occur once Consequences ensure that the desired behavior continues to occur.
Consequences are generally of 3 dimensions:

- Immediate (I) or Future (**F**)
- Certain (C) or uncertain (**U**)
- Positive (P) or Negative (**N**)

Therefore, the choice becomes one of Immediate, Certain and Positive (ICP) versus Future, Uncertain, Negative (**FUN**). So, if we are to succeed in changing the culture, we have to take the **FUN** out of Process Safety. Great! So how do we do that?

We, as process safety professionals, have to change our way of thinking. Let me introduce you to another one of my favorite authors, E.F. Schumacher. Schumacher was a German ex-pat living in England, who wrote one of the most influential books of the 20th century; "Small is Beautiful; Economics as if People Mattered." (4) It's still in print nearly 40 years after its publication. In his work Schumacher makes use of the concept of two types of thinking and problem-solving, as defined by Liam Hudson, a British psychologist.

One is called "convergent" thinking, in which the person is good at bringing material from a variety of sources to bear on a problem, in such a way as to produce the "correct" answer. This kind of thinking is particularly appropriate in science, math and technology, the kind of thinking that we, as engineers and scientists excel at.

The other is termed "divergent" thinking. Here the student's skill is in broadly creative elaboration of ideas prompted by a stimulus, and is more suited to artistic pursuits and study in the humanities.

So, if we are to succeed in taking the FUN out of process safety, we have to **THINK LIKE AN ENGLISH MAJOR—DIVERGENTLY!**

How might we begin?

One is by taking a different look at the familiar, because our success may depend on things that don't directly deal with process safety. In 1947, after the end of World War II, General Dwight D. Eisenhower, Supreme Allied Commander in Europe, and future US president, was asked what weaponry was the most critical in the successful in the invasion of Europe and the defeat of Nazi Germany. His answer was, to me, surprising:

"I attribute our success to many factors, but four of the most important things were the bulldozer, the Jeep, the 6x6 truck, and the C-47 (cargo plane). And note that none of these items was ever intended for combat."

So, in the mind of the man who masterminded D-Day, his success was not dependent upon the traditional weapons of war (although they played their part, to be sure), but upon those behind-the-lines support materiel which allowed combat operations to succeed. I believe this concept has application in the process safety arena.

We must also address each of the FUN elements in turn:

1. Turning Future into Immediate

One idea here could be to tie process safety to reliability. After all, do we really care what it's called, as long as primary containment is maintained? In doing this, what we are effectively doing is going after the higher frequency, lower risk events aggressively.

This can have immediate benefits for the bottom line:

a. Higher productivity
b. Less downtime
c. Fewer environmental events

Kind of like using a Jeep instead of a tank to win the war (see above).

2. Turning Uncertain into Certain

Here is an arena where better metrics—both leading and lagging, are critical to long term success. Fortunately, technology has improved to the point where we can now begin to get at data that was previously unavailable to us. For example, we can use of distributed control system (DCS) data to capture and analyze much more process data than we do now. However, the ability to capture data is only part of the battle. The real issue is the development of innovative, meaningful measures, so that we can make sense of what we are capturing. To quote a baseball analogy, this is the process equivalent of runs produced, the real measure of baseball success, versus RBI or batting average statistics.

3. Turning Negative into Positive

How about combining the previous two ideas; tie process safety to reliability and its immediate benefit on the bottom line, and use better metrics—both leading and lagging-to make the case. Development of innovative, meaningful measures will have a stronger influence on senior managers than anything else that we can do. Executives are data driven people!

In their March 10, 2012 issue, *The Economist* made this statement in their article on "Nuclear Power: The Dream that Failed":

"Safety requires more than good engineering. It takes independent regulation, and a meticulous, self-critical safety culture that endlessly searches for risks it might have missed."

Returning to the OECD Document on Corporate Governance, the following statement is made about High reliability organizations (HROs):

A high reliability organization has been defined as one that produces product relatively error-free over a long period of time. Two key attributes of high reliability organizations are that they:

1. ***Have a chronic sense of unease**, i.e. they lack any sense of complacency. For example, they do not assume that because they have not had an incident for ten years, one won't happen imminently.*

2. ***Make strong responses to weak signals**, i.e. they set their threshold for intervening very low. If something doesn't seem right, they are very likely to stop operations and investigate. This means they accept a much higher level of 'false alarms' than is common in the process industries.*

Will Rogers said:

"A cat that has sat on a hot stove will never sit on another hot stove—it will also never sit on a cold stove."

If we are to make **appropriate** strong responses to weak signals, they have to be weak signals that point us in the right direction. Otherwise, we could be endlessly searching (using the Economist's term) in all the wrong places. To me, this simply reinforces the need for reliable, believable, actionable metrics.

Milton Friedman, the great Chicago economist, said this about change:

"Only a crisis - actual or perceived - produces real change. When that crisis occurs, the actions that are taken depend on the ideas that are already lying around. It is worth discussing radical changes, not in the expectation that they will be adopted promptly but for two other reasons:

> *One is to construct an ideal goal, so that incremental changes can be judged by whether they move the institutional structure toward or away from that ideal."*
>
> *The other reason is very different. It is so that if a crisis requiring or facilitating radical change does arise, alternatives will be available that have been carefully developed and fully explored."*

Summary and Conclusions

In Summary, we must:

1. Think divergently—there are many paths to success, and we certainly have not found them all.
2. Set an ideal goal, and test everything you do against it to see if it is moving you in the right direction.
3. Be prepared for the crisis, so that you know what alternatives exist, and how you would choose to implement them.
4. Work to develop the self-critical structure that continues to seek out and manage risk. Accurate, believable metrics are one key to success—you have to be looking for the needle in the correct haystack!

Why? So that we can take the FUN out of process safety, that's why!

References

1. Jared Diamond, "Guns, Germs and Steel: The Fates of Human Societies" W. W. Norton & Company (1999)
2. "Corporate Governance for Process Safety: Guidance for Senior Leaders in High Hazard Industries", Organization for Economic Cooperation and Development (2012)
3. Aubrey Daniels International www.aubreydaniels.com
4. E.F. Schumacher, "Small is Beautiful: Economics as if People Mattered" Blond Briggs Publishing (1974)

(605c) Beware the Black Swan

John F. Murphy, Retired U. S. Chemical Safety Board & the Dow Chemical Company, Punta Gorda, FL

In 16th century London, a statement that describes impossibility was referred to as a "black swan" because all historical records of swans up to that time had reported that swans only had white feathers. The Greek philosopher Aristotle first used the term "white swan" as an example of necessary relations and the term "black swan" as the expression of the improbable. Nassim Nicholas Taleb in his recent book entitled *The Black Swan, The Impact of the Highly Improbable* discusses the extreme impact of rare and unpredictable events (black swans) and the human tendency to find simplistic explanations for these events after they occur. Those of us that attempt to evaluate risk must understand that black swan events cannot be predicted using current risk analysis tools. We must also remember the lessons learned from previous black swan events when designing and operating new chemical processing facilities.

This paper will discuss the concept of the black swan as applied to chemical process risk assessment. Lessons learned from black swan events of the past and the applicability of these lessons to the design and operations of existing and new chemical processing facilities will be discussed. The limitations of LOPA and other risk analysis tools in predicting blacks swan events will be examined. The discussion on the limitations of risk analysis will include examples of hazards scenarios where risk analysis tools such as LOPA may not be appropriate.

Technology Options for Coping with Beyond Design Basis Events in the Nuclear Power Generation Industry

AIChE Annual Meeting
Pittsburgh, PA
October 29, 2012

Ed Lahoda
Westinghouse Electric Company LLC

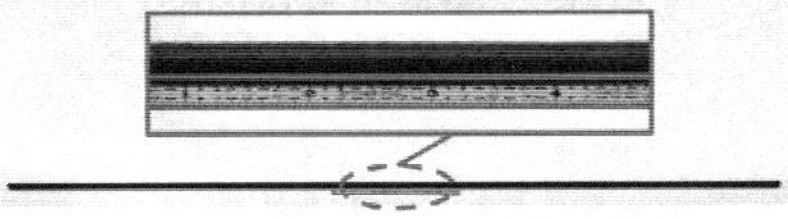

SilCar™ SiC Cladding

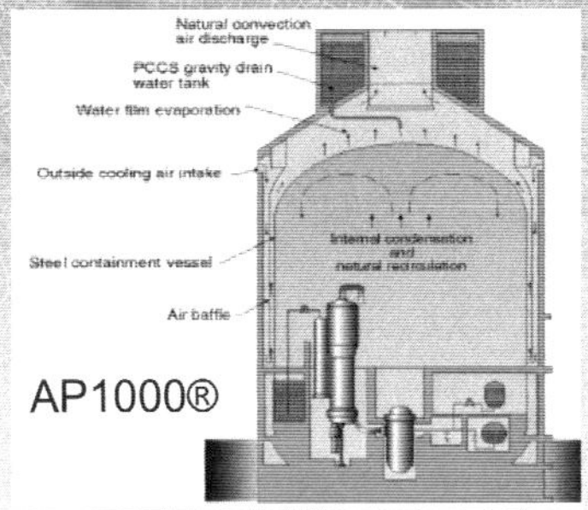

Westinghouse SMR

Station Blackout (SBO) – An Extreme Beyond Design Basis Event

Per 10 CFR 50.2 "The complete loss of alternating current (AC) electric *power to the essential and nonessential switchgear buses in a nuclear power plant (i.e., loss of offsite electric power system concurrent with turbine trip and unavailability of the onsite emergency AC power system). Station blackout does not include the loss of available AC power to buses fed by station batteries through inverters or by alternate AC sources as defined in this section, nor does it assume a concurrent single failure or design basis accident…."*

Safety functions required to mitigate SBO

1. Core cooling, inventory, and reactivity control
2. Containment Cooling and ultimate heat sink
3. Spent Fuel Pool cooling

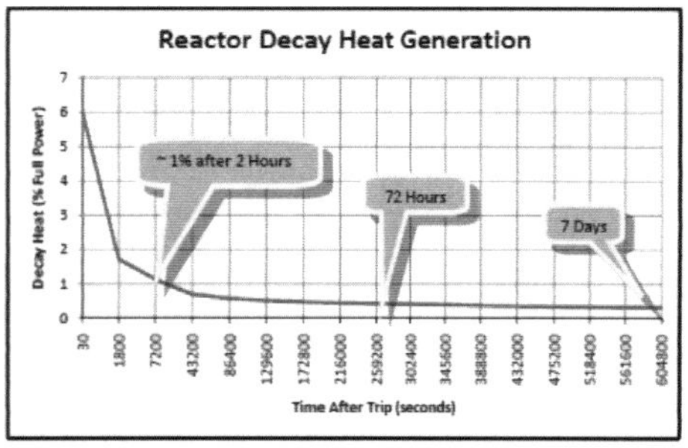

Figure 1: AP1000 Decay Heat Curve 0-7 Days

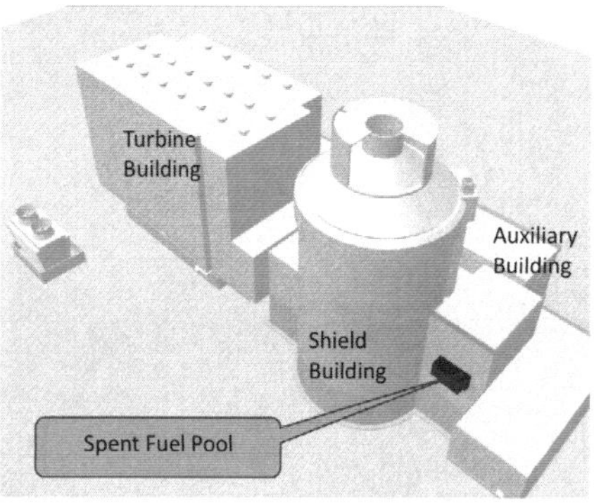

Westinghouse

Westinghouse Has Three Passive Options to Mitigate Extreme Beyond Design Basis Events

- New plants (Gen III+) being built now – AP1000®
- New generation plants in the future (Gen III+) – Westinghouse SMR (225MWe), AP1500 and other new passive plant designs
- Old plants built over last 60 years and some new plants (Gen II and Gen III) – **SilCar**™ SiC cladding

Westinghouse

Westinghouse Non-Proprietary Class 3

The AP1000 Designed To Cope With Station Blackout

- Safety-related passive systems and equipment sufficient to automatically establish and maintain Safe Shutdown Conditions for 72 hours – without operator action and without onsite and offsite AC power sources
- Includes both onsite equipment and safety-related connections for use with transportable equipment and supplies to ensure safety functions are maintained up to and beyond 7 days

Westinghouse

5

Westinghouse Non-Proprietary Class 3

Decay Heat Removal is Provided by Passive Systems for the first 3 days (72 hours)

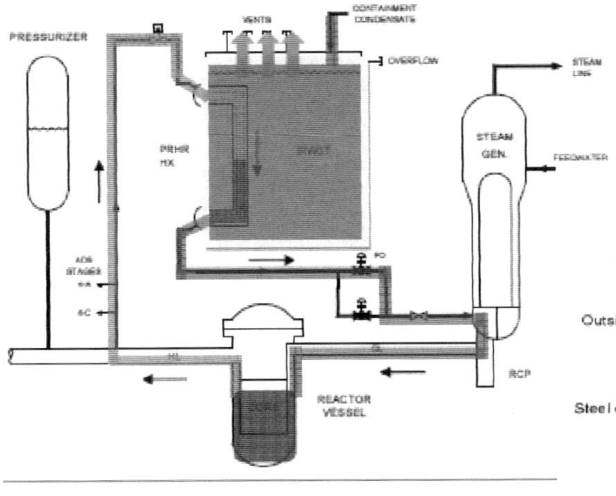

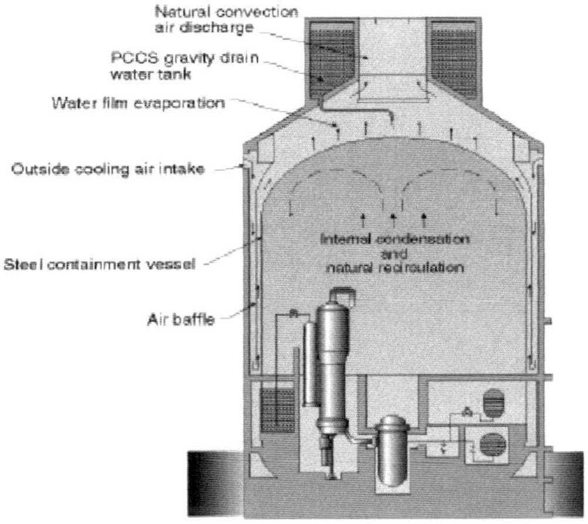

Westinghouse

6

Initiation of Natural Circulation & Decay Heat Removal

0 – 72 hours

- SBO results in reactor trip
- Remove decay heat
- Passive core cooling
- Passive containment cooling

3 – 7 days

- Start Ancillary Diesel Generators
- Provide power for control room lighting, instrumentation for plant operators, and Passive Containment Cooling Recirculation Pumps
- Passive Containment Cooling Ancillary Water Storage Tank

7 days & beyond

- Additional sources of makeup water available onsite to continue operation of Containment Cooling indefinitely
- External water sources (tanker trucks, emergency vehicles) can be used to transport fresh water

Westinghouse

Spent Fuel Pool Cooling and Makeup

0 – 72 hours

- Loss of all AC power
- Boiling occurs
- Makeup water provided to maintain level above assemblies
- Building relief panel opens automatically

3 – 7 days

- One of two PCS Recirculation Pumps (powered by the Ancillary Diesel Generators) pumps water from Ancillary Tank

7 days & beyond

- Alternative sources of water supply may be used, e.g., Fire Water Tanks and Demineralized Water Storage Tanks
- Site-specific, temporary equipment will be used to transfer water from alternative sources

Westinghouse

Next Generation Small (225MWe) Plants – Westinghouse SMR

- Compact, high-pressure containment
- Passive cooling design based on AP1000®concepts
- Targeting operation at vacuum:
  - Eliminate severe accident hydrogen burn issues
  - Increase heat transfer inside containment
  - Reduce insulation needs
  - Reduce containment overpressure challenges
- In-vessel retention strategy for severe accidents is accomplished by flooding lower containment vessel
  - Derivative of **AP1000®** design

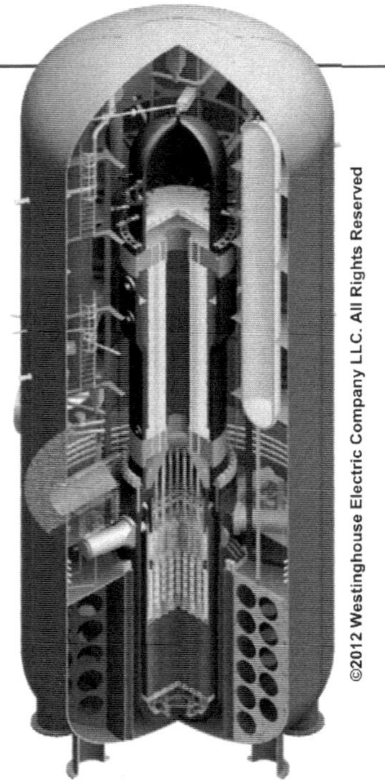

Westinghouse

Safety Video

Westinghouse

SMR Plant Layout

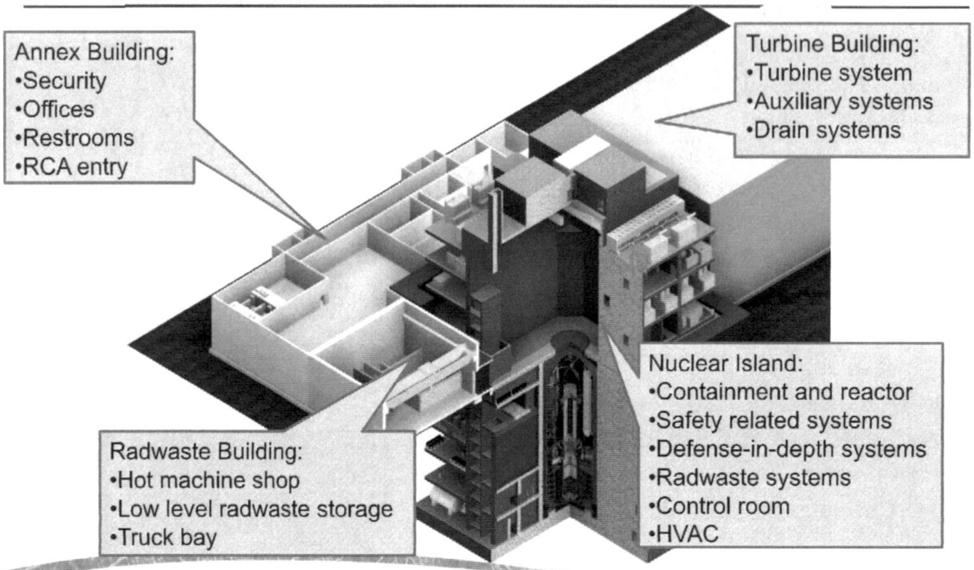

Westinghouse

Westinghouse SMR Safety Systems

- The Westinghouse SMR has over 52 plant systems
- Focus on Safety-related / defense-in-depth systems:
  - Reactor Coolant System RCS
  - Steam Generator System SGS
  - Passive Core Cooling System PXS
  - Ultimate Heat Sink System UHS
  - Containment Vacuum Filtration System VFS
  - Central Chilled Water System VWS
  - Main Control Room Emergency Habitability System VES
  - Spent Fuel Pool Cooling System SFS
  - Compressed and Instrument Air System CAS
  - Protection and Safety Monitoring System PMS
  - Class 1E dc and Uninterruptible Power Supply System IDS

Westinghouse

Generation II and III Plants – Active Safety Systems to Maintain Fission Product Boundary

- Safety systems built into plant to respond to defined set of events
  - Station blackout is considered a beyond design basis event
- Assumes that power station survives initiating event
- Assumes that either grid or emergency diesel power available for long term
- Assumes correct and prompt intervention of operating staff
- Upgrades to current plants over the years have incorporated more beyond design basis events
  - Terrorist attacks
  - Some station blackout scenarios

Westinghouse

Potential Accident Scenarios – Maximum Temperature of Core Calculations

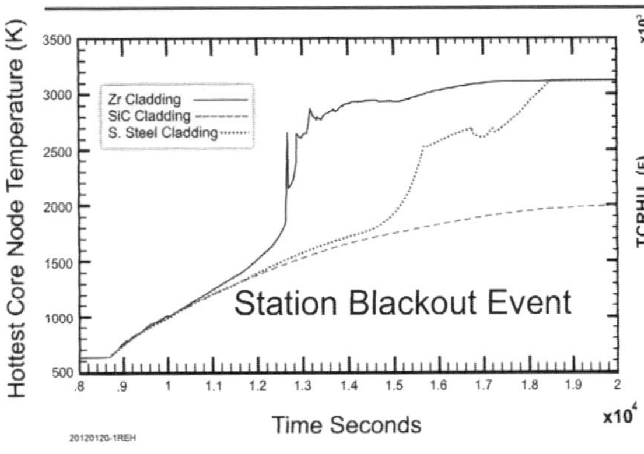

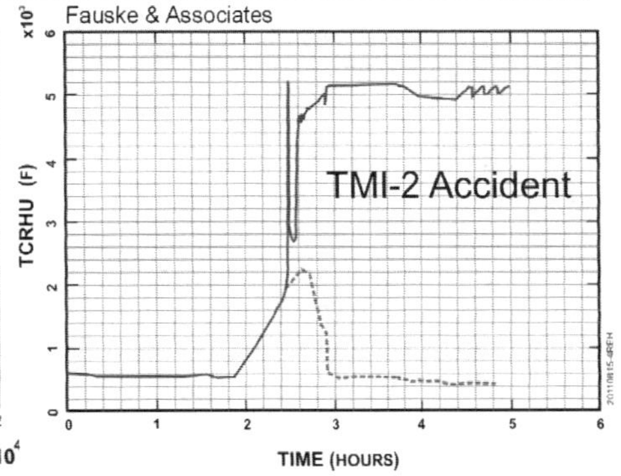

- Calculated peak core temperature.
  - 2830°C for a Zircaloy 2 clad core due to heat liberated from Zr or SS cladding oxidation.
  - 1170°C for TMI2 and ~1700°C for a non-reactive cladding core with only decay heat

Objective is to maintain fission barrier integrity during TMI 2 and Fukushima type events

Westinghouse

Westinghouse SiC/SiC Cladding Developed to Maintain Fission Product Boundary to >1700°C

SilCar™ duplex SiC LWR cladding

- Inner monolithic dense SiC inner layer (12-20 mils)
- Outer SiC/SiC composite outer layer (12-20 mils)

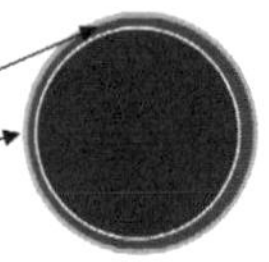

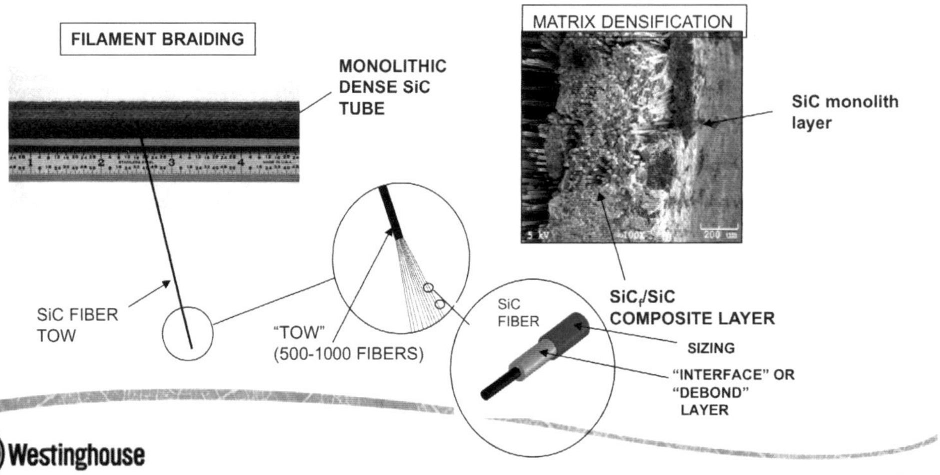

Westinghouse

SiC/SiC Structural Composites Tolerate High Burnups and High Temperatures and Reduce Water Reaction Rate

- **Unparalleled high temperature strength and low H_2 generation (reaction with steam) rates at LOCA and DNB conditions (~1200° C)**
- **Resistant to fretting and debris wear – single biggest reasons for current fuel failures**
- **Maintains strength with very high exposures – higher burnup limit lowers costs with new fuels**
- **Low Tritium permeability – better fission product barrier and allows H_2 fill to lower fuel temperature**
- **Low neutron absorption (25% of Zr alloys) – better fuel economy**
- **Fibers and monoliths available**
- **Ability to deposit ... ursors available**

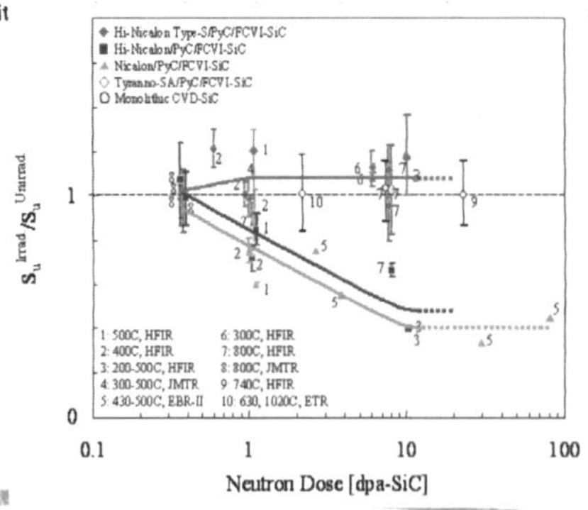

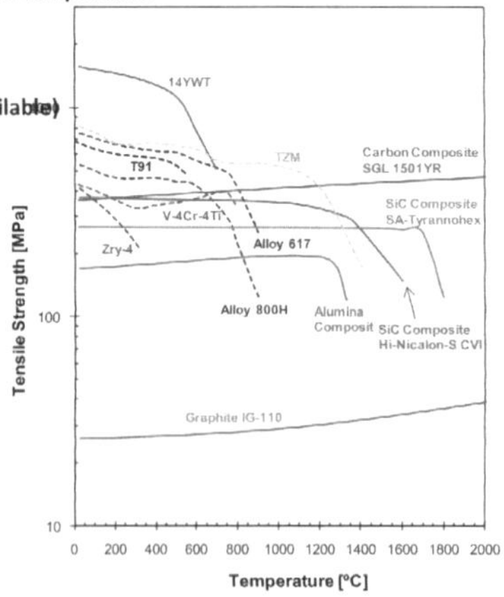

Westinghouse

Zr Alloy Coatings Also Being Explored for High Performance and Increased Safety

- Coatings of Ti_2AlC with oxidation resistance to >1400°C and thermal stability to much higher temperatures
- Attractive features
  - Lower development cost than SiC composites
  - Shorter development cycle
- Issues
  - Many potential issues need to be addressed by in-reactor testing such as response to cladding shrinkage and growth, resistance to peeling, etc.
  - Minor economic advantages

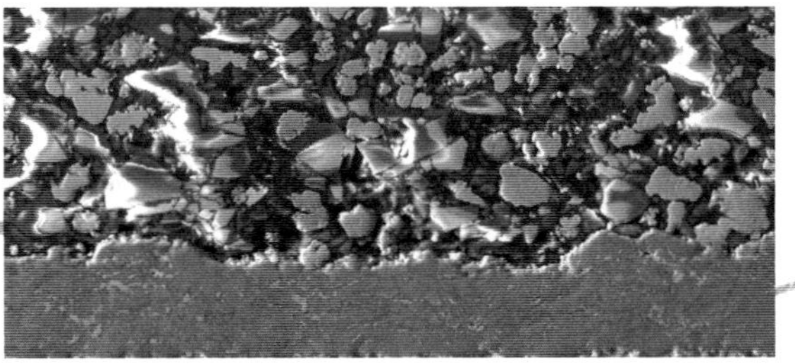

Westinghouse

Westinghouse Non-Proprietary Class 3

Westinghouse Has Been Developing **SilCar**^{TM} SiC Cladding for Over 7 years – Latest Developments

- Eight 3 ft. thin-walled monolithic SiC tubes fabricated
- Seven samples SiC-fiber wound; sectioned for destructive mechanical properties determination
- End plug sealing of composite clad tube methods have been selected and testing is progressing (3 successful so far)
- Irradiation testing in High Flux Isotope Reactor at ORNL with fuel material inside ongoing

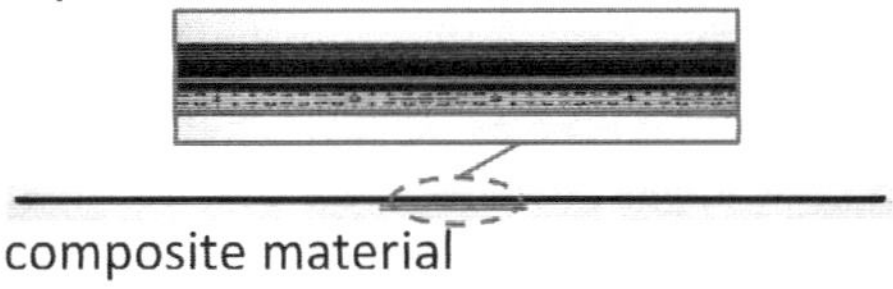

- Over next year
  - Obtain mechanical property data on composite material
  - Develop extruded closed end for long thin-walled SiC tube
  - Demonstrate SiC to SiC end plug bond on tube

Westinghouse

18

Summary – Westinghouse Actively Addressing "Black Swan" Issues While Improving Economics

- Simplified, passive designs currently available and being developed that address SBO issues while maintaining economics of nuclear power
  - AP1000 currently available (10 under construction)
  - AP1500 and SMR under development
- New claddings being developed for Generation II and III reactors with active safety designs
  - **SilCar™** SiC cladding
  - Coated Zr alloy
- High density and high thermal conductivity fuels to increase for better economics

Westinghouse

(669a) Multi-Domain Mixed Potential Model for Spent Fuel Dissolution

Kurt Frey and James L. Jerden Jr., Chemical Sciences and Engineering, Argonne National Laboratory, Argonne, IL

Rates of spent nuclear fuel corrosion and dissolution in an aqueous environment are controlled by the galvanic potential of the fuel surface. Radiolytic products present in the environment affect this potential and can enhance fuel degradation through oxidative means. However, varying degrees of fuel utilization give rise to secondary domains on the fuel matrix surface, such as micro- to nano- sized epsilon phases comprised of molybdenum, ruthenium, and various other noble metals. These secondary domains can maintain different potentials from the bulk fuel surface, and participate in reactions (specifically hydrogen oxidation) that compete with other fuel dissolution reactions. This work extends the mixed-potential model (MPM) developed by Shoesmith, Kolar, and King to incorporate additional fuel surface domains. Originally used to describe the oxidative dissolution of uranium dioxide fuel in the presence of radiolytically generated hydrogen peroxide, the additional MPM domains have been used to model the epsilon phase and its impact on the rates of spent fuel dissolution.

(669b) Plant-Level Modeling and Simulation of Fuel Recycling Processes: Commonality and Integration

Valmor F. de Almeida, Energy and Environmental Sciences Directorate, Oak Ridge National Laboratory, Oak Ridge, TN

Plant-Level Modeling and Simulation of Fuel Recycling Processes: Commonality and Integration

Valmor F. de Almeida

Oak Ridge National Laboratory, Oak Ridge, TN 37831-6181

Abstract

Used nuclear fuel recycling processes are known to be diverse and complex. This has led to a case-by-case approach for developing mathematical models and their corresponding simulation modules. This approach has many drawbacks. First, it does not provide a common theoretical framework to be applied to all processes at a well defined length and time scale. Second, as a result, it is difficult to couple mathematical models of different processes in the plant flowsheet and define a verification/validation strategy. Last, the resulting computational modules that reflect the implementation of these models, inherit the same difficulties at the algorithm and software engineering level. That is, the software realization of mathematical models are not intended to be integrated and lack a theoretical basis for coupling. In addition, computational modules do not take advantage of internal algorithms/solvers, and modules either do not have or reuse a software architecture. Given this status, modeling and simulation for fuel recycling processes at the plant level can take advantage of a modern approach taking into account commonality and integration as a basis for future developments. A desirable advantage of this approach is to provide a standard for verification, validation, and uncertainty quantification of models and modules of processes while reducing the cost of development. This scientific paradigm is needed in the nuclear recycling domain to improve robustness of simulation predictions, and gain public and political acceptance; the latter is a unique aspect of the nuclear recycling enterprise.

The work in this presentation is aimed at identifying a modeling commonality for various processes in a recycling plant using a network approach. A plant or flowsheet is modeled as a network of computational modules. These modules are designed with common external and internal behaviors to allow for integration and modular plant design. Internally, the modules result from the implementation of a model stemming from a physico-chemical network transport theory. The application of a common theory for all processes enables the development of a systematic set of modules. A central concept is the abstraction of contactors as a network of stages representing the mechanics of phase contact. This abstraction enables the development of a common structure for various separation processes wherein the application of equilibrium and non-equilibrium considerations will be independent of the particular contacting equipment. This approach allows for different kinetics and thermodynamics models to be applied to a contactor network without internal changes in the contactor model. In addition, the approach supports the development of a reduced-order model wherein subscale phenomena can be taken into account in a more systematic way. The resulting balance equations for the state variables of the network model provide the means for evolving the state of the system and for predicting quantities of practical interest. Time-coupling of modules is achieved in asynchronous way to allow for greater flexibility in developing models and integrating the resulting modules. This approach also leverages computation concurrency in parallel machines.

Computational examples of commonality and integration will be described for three processes of interest to aqueous recycling, namely, voloxidation, dissolution, and solvent extraction. These seemingly different processes share the above-mentioned commonalities and can be integrated in a more straightforward way than if they had been developed in isolation. A description of how model fidelity can be improved will be made by identifying quantities which can be provided by subscale models.

Nuclear Engineering
THE UNIVERSITY OF UTAH

Neutron Energy Spectrum Correction based on Capture Time

Can Liao
Xianfei Wen
Haori Yang, Ph.D.
Nuclear Engineering Program
University of Utah

Nuclear Engineering
THE UNIVERSITY OF UTAH

Outline

- Introduction
- Concerns
- Method
- Pulse Shape Discrimination
- Simulation
  - Analysis
  - Results
- Optical Simulation
- Experiment
- Conclusion
- Acknowledgement

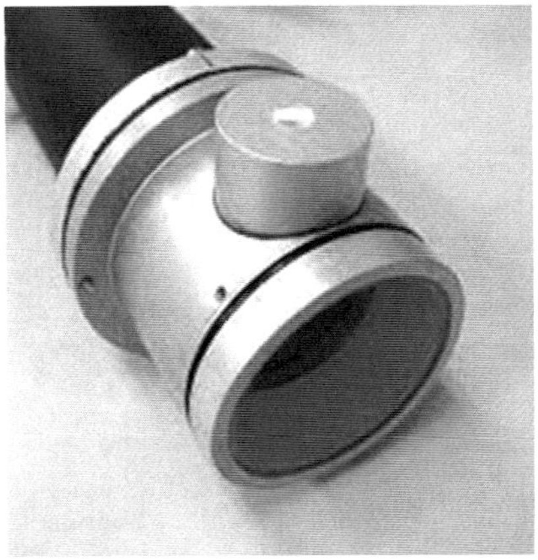

2

Nuclear Engineering
THE UNIVERSITY OF UTAH

Introduction

- Neutron Spectrometry
- Current technique
  - Time-of-flight
  - proton recoil
  - spectrum de-convolution
- Capture-gated method

  ----Fast neutrons

3

Nuclear Engineering
THE UNIVERSITY OF UTAH

Introduction

- Fast neutron enters the detector

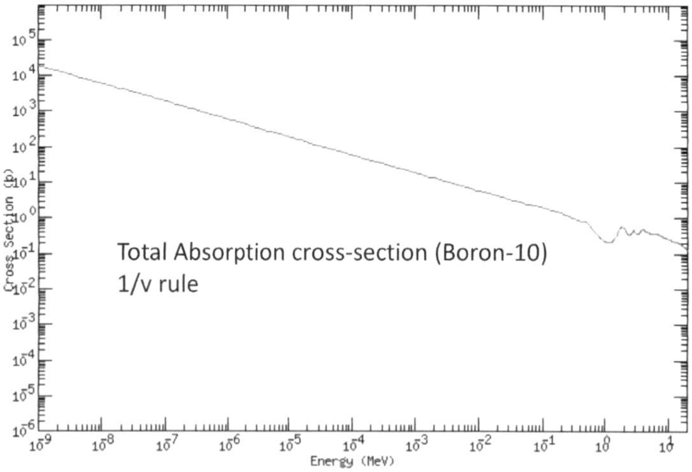

4

Nuclear Engineering
THE UNIVERSITY OF UTAH

Introduction

- Fast neutron enters the detector
- Two Pulses
  - Recoil proton Pulse

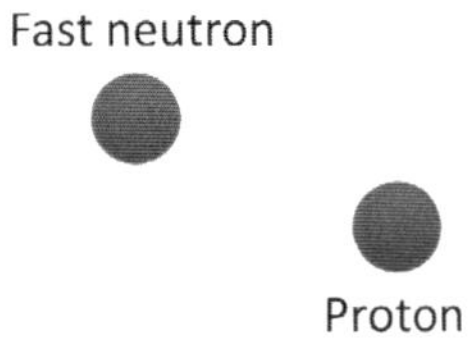

5

Nuclear Engineering
THE UNIVERSITY OF UTAH

Introduction

- Fast neutron enters the detector
- Two Pulses
  - Recoil proton Pulse

6

Nuclear Engineering
THE UNIVERSITY OF UTAH

Introduction

- Fast neutron enters the detector
- Two Pulses
  - Recoil proton Pulse

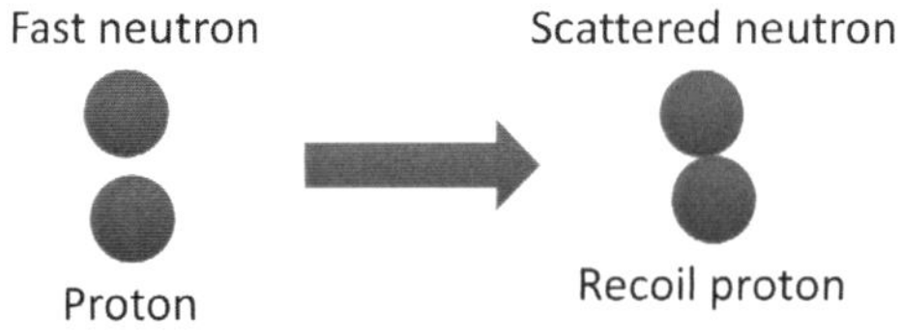

7

Nuclear Engineering
THE UNIVERSITY OF UTAH

Introduction

- Fast neutron enters the detector
- Two Pulses
  - Recoil proton Pulse

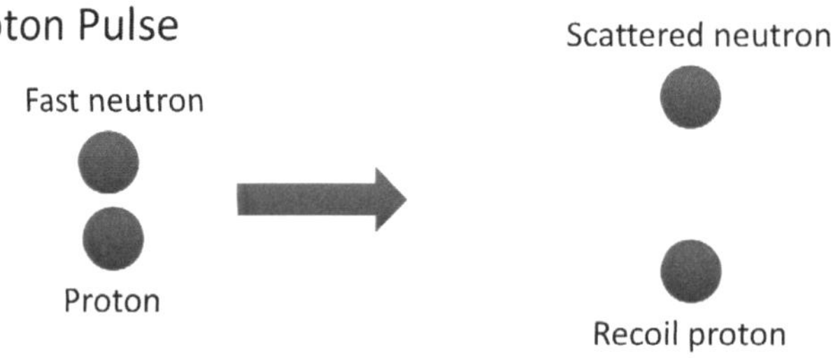

8

Nuclear Engineering
THE UNIVERSITY OF UTAH

Introduction

- Fast neutron enters the detector
- Two Pulses
  - Recoil proton Pulse

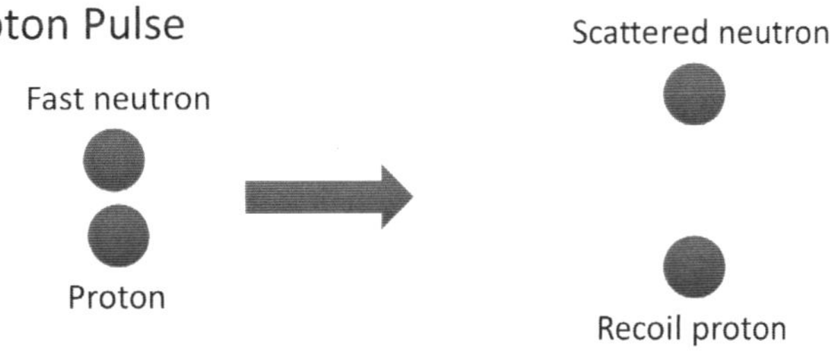

  - Neutron absorption Pulse

    neutron-absorbing medium: ^{10}B, ^{6}Li and ^{113}Cd.

9

Nuclear Engineering
THE UNIVERSITY OF UTAH

Introduction

- What is the Capture-gated method?

The scattered neutron pulse is accepted as the 'true' neutron pulse only if the captured neutron pulse is detected.

Scattered neutron pulse Amplitude

⇩

Incident neutron energy

⇩

Neutron spectrum

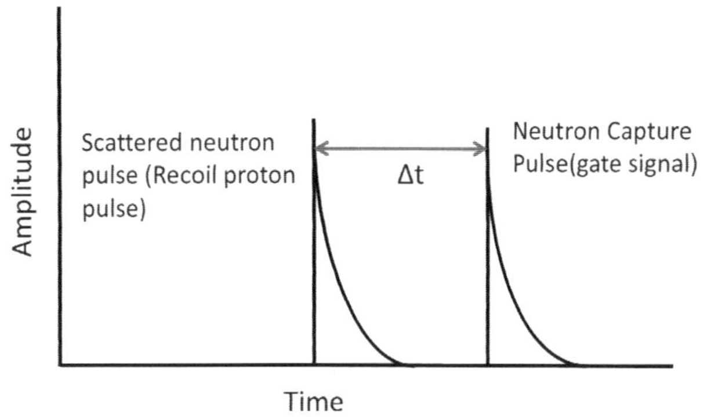

10

Nuclear Engineering
THE UNIVERSITY OF UTAH

Concerns

- The Assumption: neutron energy when captured = 0
- But this is not always true

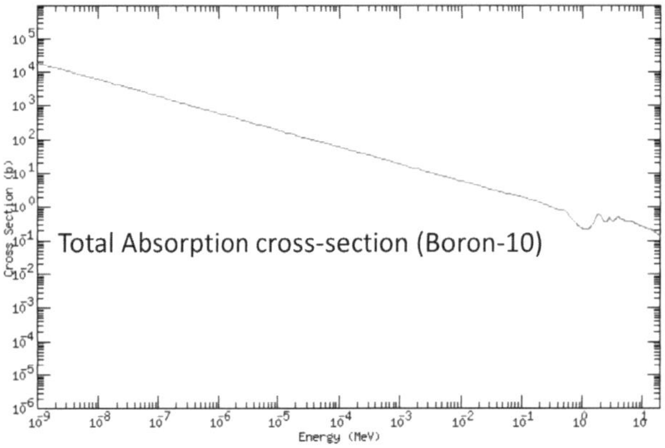

11

Nuclear Engineering
THE UNIVERSITY OF UTAH

Method

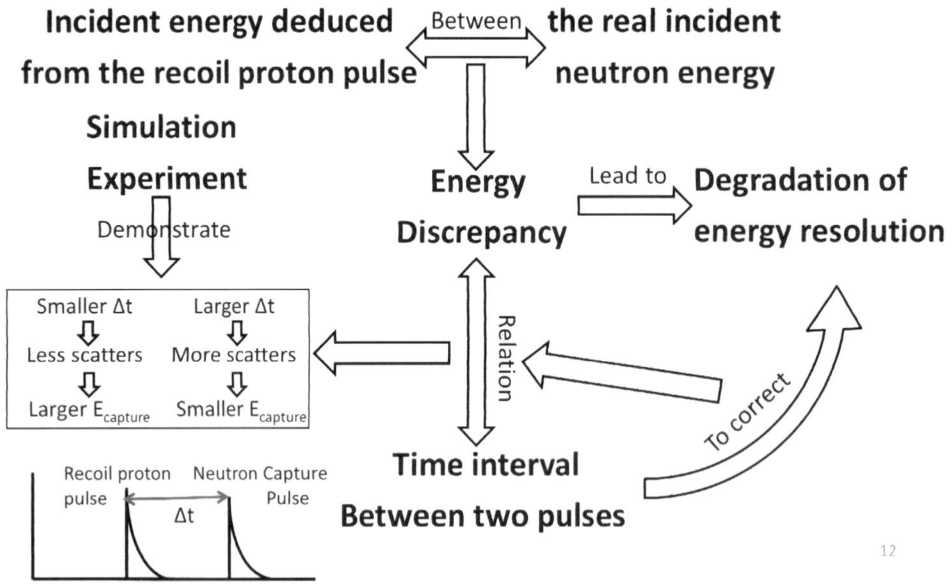

12

Pulse Shape Discrimination

- To implement capture-gated method
- Distinguish scattering, capture and gamma pulses.
- Rise-time and tail-to-total integration Algorithm

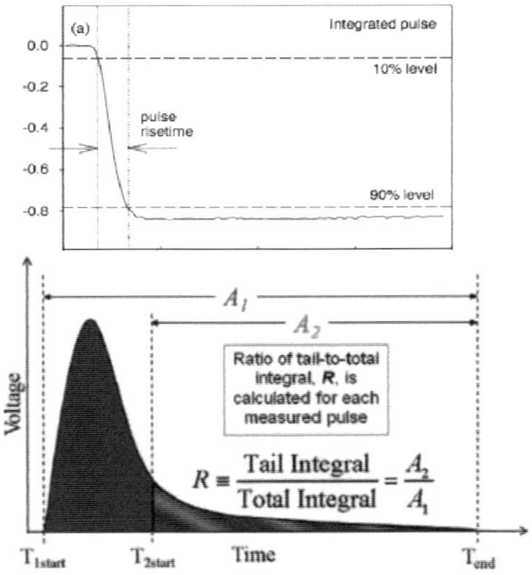

Nuclear Engineering
THE UNIVERSITY OF UTAH

Simulation

- Toolkit: Geant4
- Experiment Apparatus:
  - EJ-339A Boron Loaded Liquid Scintillator
  - Capture material: Boron-10

    $^{10}B + n \rightarrow \alpha + {}^{7}Li + \gamma$
  - Atomic Composition (atoms/cc x 10^{22})

Boron-10	0.053
Carbon	2.9
Hydrogen	5.03
Oxygen	0.814

Simulation

- 0.1 MeV monoenergetic neutron beam
- Neutron energy before captured
- Time interval
- Using Matlab to calculate and plot a 2-D Histogram for $E_{capture}$ and Δt

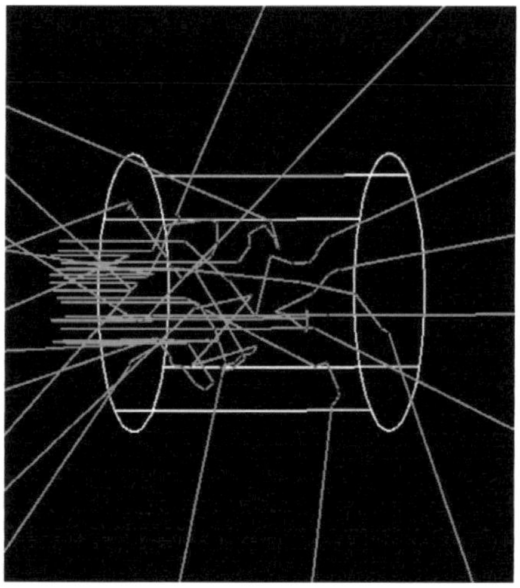

15

Nuclear Engineering
THE UNIVERSITY OF UTAH

Simulation

Analysis

- X-axis: $\log_{10}(\Delta t)$
- Y-axis: $\log_{10}(E_{capture})$

Smaller Δt	Larger Δt
⇩	⇩
Less scatters	More scatters
⇩	⇩
Larger $E_{capture}$	Smaller $E_{capture}$

- For a particular Δt, what about E?

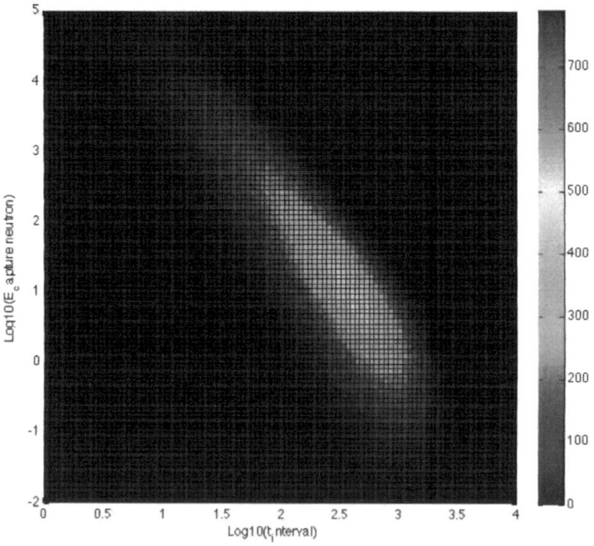

16

Simulation

Analysis

- Slice at $\log_{10}(\Delta t) = 2.5$
- For a particular Δt, $\log_{10}E$ follows a certain distribution
- $E_{capture}$ is the needed Energy Discrepancy
- Derive a mathematical relation between $E_{capture}$ and Δt
- i.e. Drive a Energy distribution for this Δt

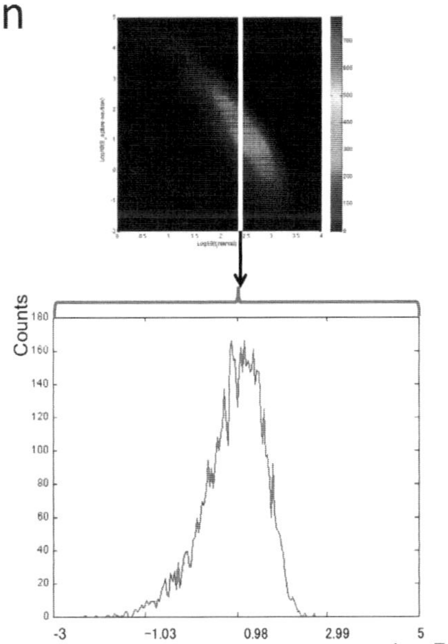

17

Simulation

Analysis

- Then correct the discrepancy at this Δt according to the distribution
- Then do this similarly at other Δts
- Get the Energy correction for all Δts

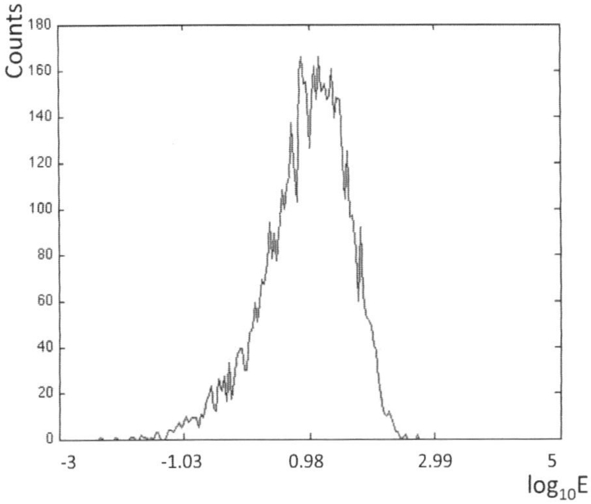

18

Simulation

Result

Corrected spectrum

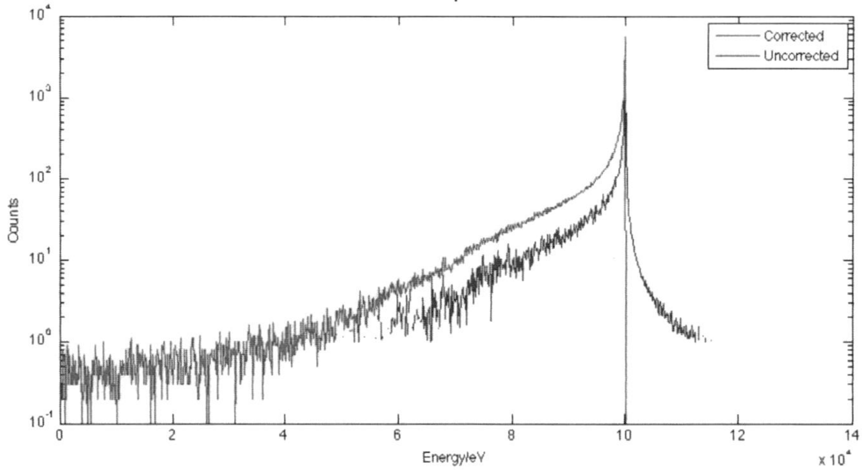

19

Nuclear Engineering
THE UNIVERSITY OF UTAH

Optical Simulation

- Add Optical process into simulation

G4double naienergies[numentriesnai] = { 2.48*eV, 2.58*eV, 2.69*eV, 2.75*eV, 2.82*eV, 2.85*eV, 2.88*eV, 2.92*eV, 2.94*eV, 2.95*eV, 3.10*eV }; // max: 425 nm

G4double naifastcomp[numentriesnai] = { 0.05, 0.18, 0.44, 0.53, 0.75, 0.85, 0.95, 1.0, 0.93, 0.88, 0.0 };

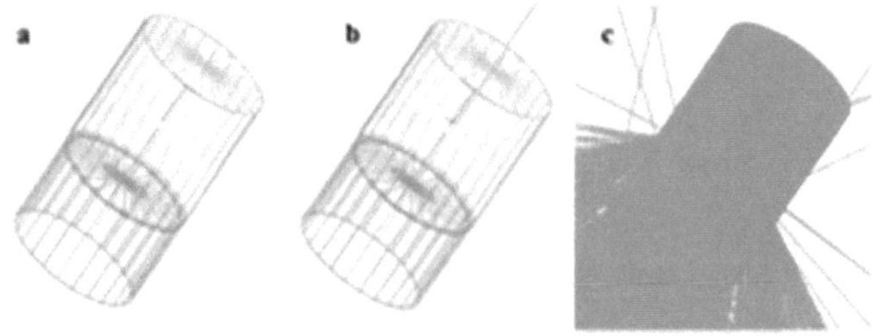

20

Optical Simulation

- Record time of every optical photon
- Histogram times of optical photons
- The shape of Pulse before shaping
- Implement the tail-to-total algorithm

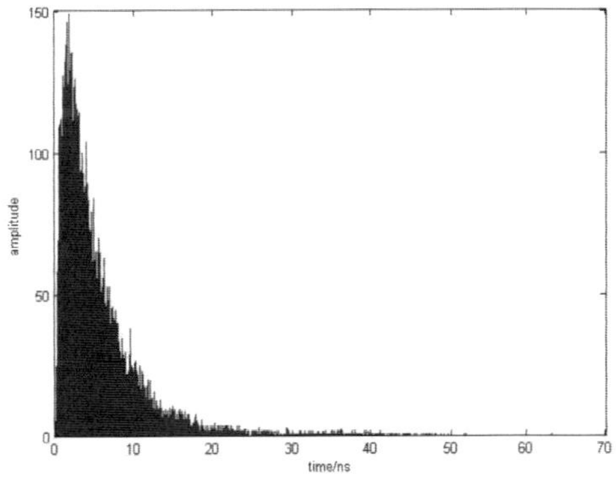

Nuclear Engineering
THE UNIVERSITY OF UTAH

Optical Simulation

Result

To be added

Nuclear Engineering
THE UNIVERSITY OF UTAH

Experiments

- Using Cs-137 γ source（0.662MeV）
- Using Labview and Matlab to record data
- Pulse after shaping
- Rise-time distribution

  rise-times of gamma pulses are all round 10.2s

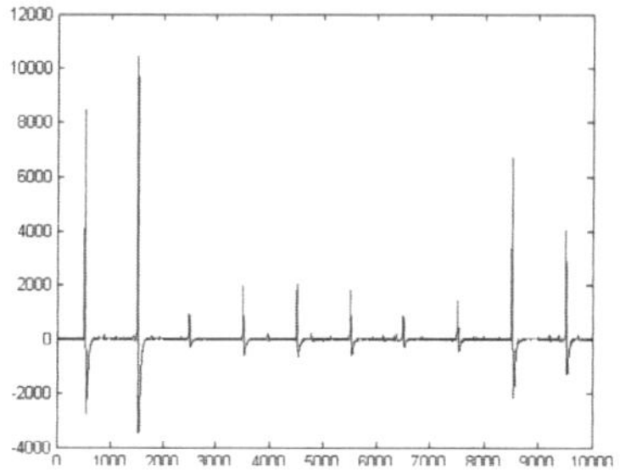

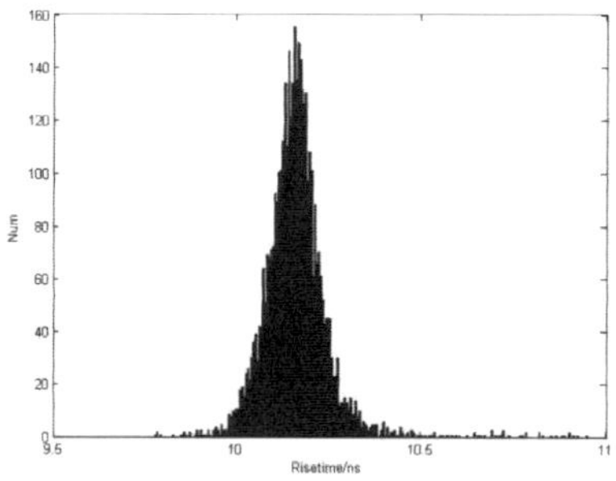

3

Nuclear Engineering
THE UNIVERSITY OF UTAH

Conclusion and Future work

- Spectrum correction based on capture time is feasible
- EJ339A has good PSD features
- However, a more accurate correction function is necessary
- Simulating more particles to obtain a more realistic result
- More Experiments

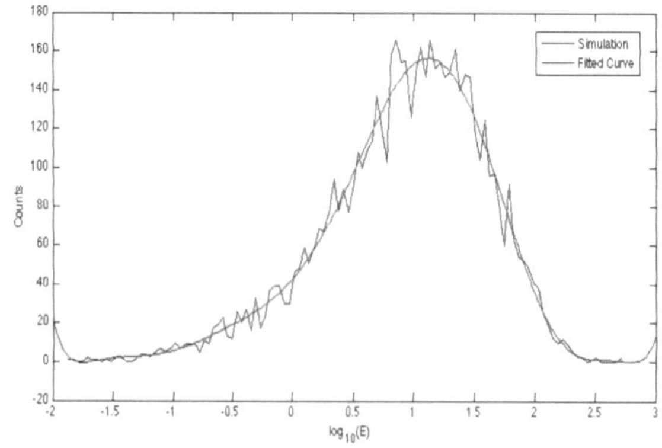

24

Nuclear Engineering
THE UNIVERSITY OF UTAH

Acknowledgement

Haori Yang, PhD, Assistant Professor, Nuclear Engineering Program, Univ. of Utah

Dong-Ok Choe, PhD, Assistant Research Professor, TRIGA reactor supervisor, Nuclear Engineering Program, Univ. of Utah

Tatjana Jevremovic, PhD, Professor, Director of the Nuclear Engineering Program, Univ. of Utah

25

Nuclear Engineering
THE UNIVERSITY OF UTAH

Thank you!

26

EVALUATION OF THE INTEGRATED SOLUBILITY MODEL, A GRADED APPROACH FOR PREDICTING PHASE DISTRIBUTION IN HANFORD TANK WASTE

Kayla L. Pierson, Jeremy D. Belsher and Kendra R. Seniow, Washington River Protection Solutions, Richland, WA

Background and Introduction

Approximately 56 million gallons of radioactive and chemically hazardous waste is stored in 177 underground storage tanks at the Department of Energy (DOE) Hanford Site in southeastern Washington. The mission of the DOE River Protection Project (RPP) is to store, retrieve, treat and dispose of Hanford's tank waste. Waste is retrieved from the underground tanks and delivered to the Waste Treatment and Immobilization Plant (WTP). Waste is processed through a pretreatment facility where it is separated into low activity waste (LAW), which is primarily liquid, and high level waste (HLW), which is primarily solid. The LAW and HLW are sent to two different vitrification facilities and glass canisters are then disposed of on-site (for LAW) or shipped off-site (for HLW). The RPP mission is modeled by the Hanford Tank Waste Operations Simulator (HTWOS), a dynamic flowsheet simulator and mass balance model that is used for mission analysis and strategic planning.

HTWOS tracks and predicts the phase distribution of over 130 radioactive and non-radioactive components in tank waste. Solubility modeling in HTWOS is not thermodynamically based and waste phase distributions are approximated by zero-order dissolution split factors that are based on extrapolation from limited experimental data. Component solubility in HTWOS is very important for mission modeling as component phase distributions impact the quantity of each component reporting to LAW or HLW. Both waste glass types must meet a variety of processing, acceptability and model validity constraints. Any constraint that limits the amount of waste that can be incorporated into the waste glass is referred to as a "glass driver." Small changes in phase distribution can have large impacts on the quantities of waste glass produced, the mission duration, and the lifecycle cost.

The HTWOS Sensitivity Study (RPP-RPT-51819, Rev. 0) evaluated the impact that varying 21 different parameters in HTWOS had on selected mission metrics, i.e. glass canisters, mission length. The parameters were grouped into four categories: inventory, solubility, retrieval duration and WTP. The solubility parameters had the greatest overall impact of all the categories. The phase partitioning of sulfate is extremely significant as it is a primary glass driver for HLW with a very limited quantity being incorporated into the glass. In the current HTWOS framework sulfate is very soluble, thus reporting primarily to LAW. If sulfate is less soluble than predicted by the current split factors, more reports to HLW, increasing the canister count and extending the mission. Aluminum and phosphate solubility also have a large impact on the canister count and mission duration, although not to the same extent as sulfate.

The integrated solubility model (ISM) was developed to improve the chemistry basis in HTWOS and better predict the outcome of the RPP mission. The ISM uses a graded approach to focus on the components that have the greatest impact to the mission while building the infrastructure for continued future improvement and expansion. Components in the ISM are grouped depending upon their relative solubility and impact to the RPP mission. The solubility of each group of components is characterized by sub-models of varying levels of complexity, ranging from simplified correlations to a set of Pitzer equations used for the minimization of Gibbs Energy. The four different categories are described in detail below.

Category 1 components are very soluble or very insoluble and have a low impact to the RPP mission. Each of these components is assumed to reside exclusively in either the liquid or solid phase. The logic and strategy for binning components is described in RPP-RPT-47542, Rev. 0, *Simplified Solubility Estimates for Selected Waste Compounds*.

Category 2 components exhibit intermediate solubility and have a low impact to the RPP mission. Their solubility is described by simplified equations that were developed in RPP-RPT-47542 using the Environmental Simulation Program[1] and literature data.

Category 3 components exhibit intermediate solubility and have a high impact to the RPP mission. The phase distribution for these components is determined thermodynamically using the Pitzer ion-interaction model ["Thermodynamics of Electrolytes. I. Theoretical Basis and General Equations" (Pitzer 1973) and "Thermodynamics of Electrolytes: IV. Activity and Osmotic Coefficients for Mixed Electrolytes" (Pitzer and Kim 1974)]. The Pitzer parameters were collected and made consistent for the components in Category 3 by reconciling experimental data and parameters published in literature as outlined in RPP-RPT-50703, Rev. 0, *Development of a Thermodynamic Model for HTWOS*. Solids in the Pitzer equations must be explicitly identified and were initially chosen based on RPP-PLAN-46002, Rev. 0, *Wash and Leach Factor Work Plan*, but additional solids and aqueous species were included to ensure functionality of the Pitzer equations over a wide range of conditions.

Category 4 concerns kinetic dependent species. It contains only one component, boehmite (AlOOH), which has been shown to be an inert solid under all relevant conditions except during the caustic leaching step of pretreatment. The amount of solid boehmite dissolved during caustic leaching is predicted using a kinetic dissolution equation.

Methodology and Assumptions

The ISM is being evaluated to ensure that it accurately predicts the phase distribution of Hanford tank waste. The basis and methodology for the ISM evaluation is explained in RPP-51192, Rev. 0, *Plan for the Evaluation of the HTWOS Integrated Solubility Model Predictions*, but the general approach follows. Experimental data from two types of studies was input into MS Excel®[2] to track the component masses through the precipitation and dissolution of solid species. The two study types were 1) boildowns, where liquid waste is evaporated until a large quantity of solids form, and 2) saltcake dissolutions, where solid waste is mixed with water in successive washes until a large portion of the waste dissolves. Mole values of each component in each sample were input into HTWOS and the ISM predicted the phase distribution. ISM predictions were converted to molality, compared to the experimental molality and statistically analyzed. ISM component molality predictions within a factor of 2 of the experimental molality were considered good. The solids predicted by the ISM were qualitatively compared to the

[1] Environmental Simulation Program (ESP) is a registered trademark of OLI Systems, Inc. Morris Plains, NJ. ESP is simulation software developed for electrolyte chemistry.
[2] MS Excel® is a registered trademark of Microsoft Corporation, Redmond, WA.

experimental data to determine if the ISM was predicting the correct solid, but not necessarily the correct amount.

The statistical software JMP®[3] was used to analyze the experimental and ISM molality on a component basis to find trends in the ISM predictions and identify which areas of the ISM were in most need of further development. JMP® creates a model to predict the experimental molality based on the ISM molality in the form of Equation 1. The experimental molality is M_{exp}, M_{ISM} is the ISM molality, *Tank* is the tank where the sample originated listed in JMP® as an integer value, and *A*, *B*, *C*, and *D* are coefficients. The fixed effects *Tank* and $Tank*M_{ISM}$ are included to explain potential variability in the molality predictions that are due to the differences in the tank waste data. The value of *A* for each component indicates how the ISM molality needs to be adjusted to equal the experimental molality. Values of *B*, *C* and *D* are can help explain the variance in the predictions, but the analysis will focus on the value of *A*. An *A* value of 0.5-2 is considered good.

$$M_{exp} = A(M_{ISM}) + B(Tank) + C(Tank * M_{ISM}) + D \qquad \text{Eq. 1}$$

Due to the differences in the study types and the different errors associated with each, JMP® made two separate models, one for boildowns and one for saltcake dissolutions, for each component to describe the relationship between the ISM and experimental molality.

Both study types analyzed only a small subset of the 139 components tracked by HTWOS; therefore, the evaluation will focus on the components in Category 3, the Pitzer components, with a few additional components, as listed in Table 1. Solids that will be reviewed are also listed in Table 1. Solids on this list were seen in the experimental data or were predicted by the ISM. The list does not include all the solids present in the ISM.

The evaluation focused on the evaporation and dissolution of tank waste. The evaluation did not look at how well the ISM predicted phase partitioning in the WTP.

Table 1. Components and solids analyzed in the ISM evaluation

Component	Component symbol		Solid	Solid Name
Aluminum	Al		$Al(OH)_3$	Gibbsite
Oxalate	C_2O_4		$Na_2C_2O_4$	Sodium oxalate
Chloride	Cl		Na7F(PO4)2.19H2O	Sodium fluoride phosphate
Chromium [a]	Cr		NaF	Sodium fluoride
Fluoride	F		$NaNO_3$	Sodium nitrate
Sodium	Na		$Na_3PO_4 \cdot 0.25NaOH \cdot 12H_2O$	Trisodium phosphate
Nitrite	NO_2		Na_3FSO_4	Sodium fluoride sulfate
Nitrate	NO_3		$Na_2CO_3 \cdot H_2O$	Sodium carbonate monohydrate
Hydroxide	OH			
Phosphate	PO_4			
Sulfate	SO_4			
Carbonate	CO_3			

[a] Chromium is in Category 2

[3] JMP® is a registered trademark of SAS, Cary, NC.

Results and Conclusions

Two examples of JMP® output data are shown in Figures 1 and 2 for two components: chloride on the left and chromium on the right. Figure 1 shows the regression in JMP® comparing the ISM and experimental molality for the four different tanks analyzed in boildown studies. Lines and markers correspond to specific tanks. Figure 1 shows how well the ISM predictions match the experimental data. Figure 2 compares the experimental molality to the experimental molality as predicted by the JMP® model and shows how well the JMP® model predicts the experimental molality. The red solid line represents an exact match and the more data points that are closer to the red line, the better the fit by the JMP® model. The dashed red lines represent the uncertainty in the JMP® model. The confidence in the value of *A* is dependent on how well the JMP® model predicts the experimental molality. Chloride has good predictions by the ISM as shown by the good fit of the regressions to the data. In addition, the JMP® model works well for chloride in boildowns. In contrast, chromium has poor predictions by the ISM as evidenced by the ill-fitting and irregular regressions.

Figure 1. JMP® regressions for chloride and chromium in boildown studies.

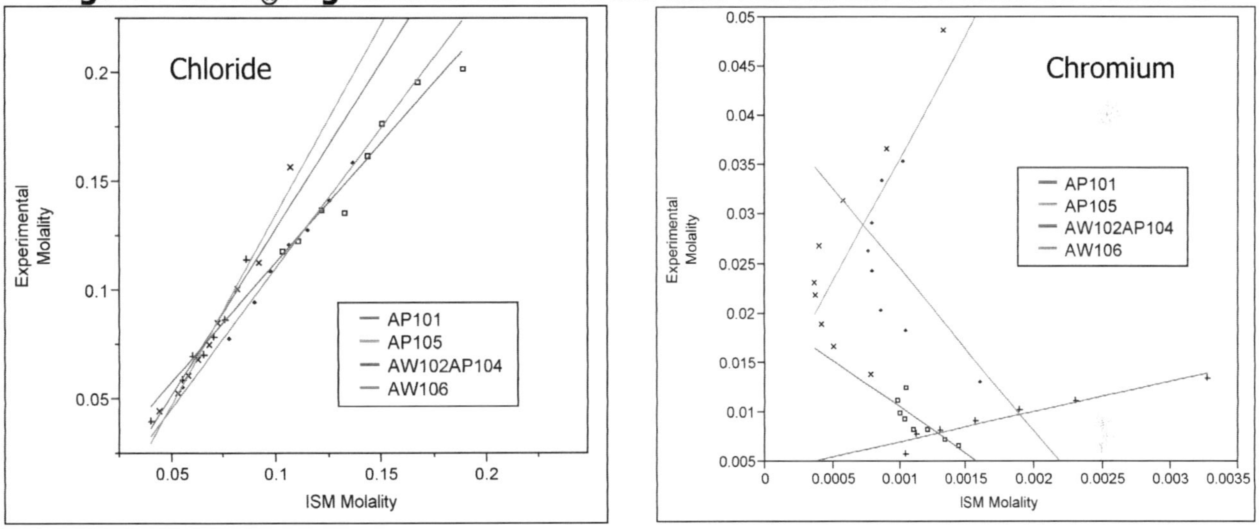

Figure 2. JMP® actual versus predicted experimental molality for Cl and Cr in boildowns.

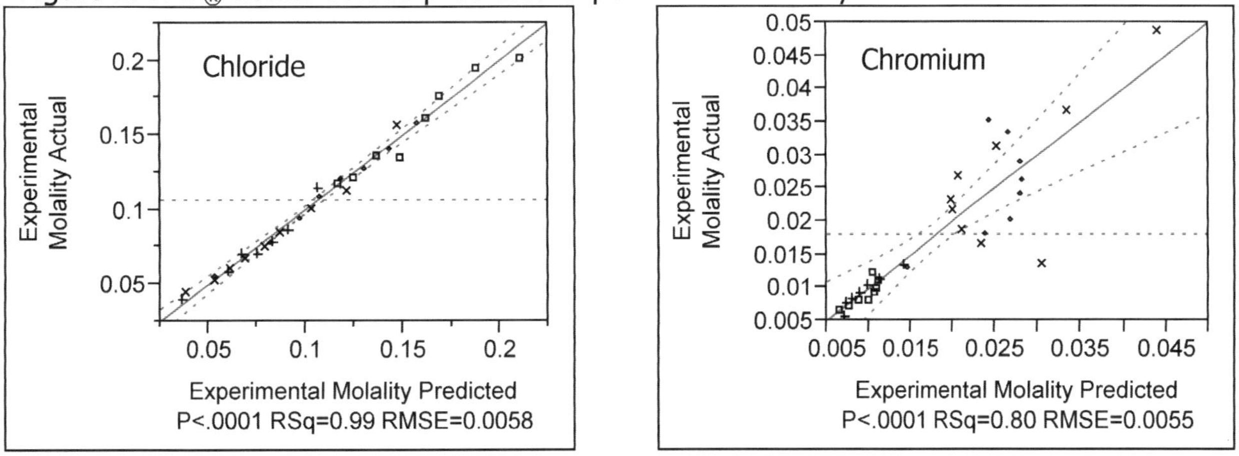

Component predictions by the ISM are considered good or poor based on the value of the coefficient *A* in the JMP® model. Table 2 lists the 95% confidence intervals of *A* for each component based on study type (boildown or saltcake dissolution). Values shaded are not good, based on the 0.5-2 criteria previously mentioned with a few exceptions.

Table 2. Ninety five percent confidence intervals for *A* based on study type.

	Boildown		Saltcake Dissolution	
Component	**Lower 95%**	**Upper 95%**	**Lower 95%**	**Upper 95%**
Al	0.77	3.51	-1.04	6.38
C2O4	-3.74	6.48	.27	1.389
Cl	0.93	1.81	0.14	1.43
Cr	-38.28	44.33	-	-
F	-0.32	1.52	-0.42	1.77
Na	0.65	1.05	0.56	1.29
NO2	0.93	1.39	0.38	1.25
NO3	0.49	1.25	0.42	1.30
OH	0.68	1.59	0.07	0.85
PO4	-1.07	1.13	0.12	1.19
SO4	0.65	1.83	-0.07	0.92
CO3	0.81	1.52	-1.11	2.37

The components that are very soluble in tank waste, i.e. Cl, NO_2, NO_3, OH and CO_3, are predicted well by the ISM in the boildowns. This is expected as the soluble components should stay in solution throughout the evaporation. Sodium is also predicted well in boildowns. There are large amounts of sodium in the waste, so small irregularities in solids precipitation (as almost all solids have sodium) will have minimal to no effect on the sodium molality. Sulfate is also fairly soluble in tank waste and occasionally forms the solid sodium fluoride sulfate (Na_3FSO_4). Sulfate is predicted well in the boildowns by the ISM and the predicted formation of Na_3FSO_4 by the ISM matches the experimental data.

Aluminum concentration is under predicted in boildowns by the ISM, meaning that the liquid concentration predicted by the ISM is lower than observed experimentally. The ISM predicts the formation of gibbsite [$Al(OH)_3$] in every sample when it is not seen in the experimental data in any sample. However, under prediction of aluminum solubility is common in tank waste solubility models ["The apparent solubility of aluminum (III) in Hanford high-level waste" (Reynolds 2012)]. The ISM has inconsistent predictions for C_2O_4. There is significant scatter in the data for oxalate and the ISM does not handle it well. The ISM does accurately predict the presence of sodium oxalate ($Na_2C_2O_4$), but it does not necessarily predict the correct quantity. The liquid concentration of fluoride and phosphate are both over predicted by the ISM due to the lack of $Na_7F(PO_4)_2 \cdot 19H_2O$ precipitation or precipitation in too small of quantity when compared to the experimental data. The chromium predictions by the ISM are poor, as evident in Figure 1.

The saltcake dissolution data was spans a wider concentration range than the boildown data, contains minimal solids analysis, and is subject to more calculation error than the boildown data. Determining why component predictions in the ISM did or did not do well was difficult or not possible based on the narrow information provided in the experimental reports. Solids analysis was performed on the initial composites but not after each wash, so comparing the ISM predictions of the specific solids dissolved to the experimental data was based on the limited inferences made in the experimental report. The composite saltcake samples were very dry and the ISM could not often make predictions until the second or third washing step due to ionic strength restrictions in the ISM. Trends in component distribution and solids analysis were challenging without ISM phase distribution and solid formation predictions in the first washes.

Only two of the five components that are very soluble in tank waste, NO_2 and NO_3, are predicted well by the ISM in saltcake dissolutions. Very soluble components should dissolve readily. Sodium is predicted well in the saltcake dissolutions for the same reasoning as in the boildowns, the large quantity of sodium in the waste makes small irregularities in solids precipitation have minimal to no effect on the concentration. The oxalate prediction for saltcake dissolutions is also good.

Aluminum and carbonate had inconsistent predictions by the ISM but the reasons for inconsistencies are unclear. The ISM over predicts the liquid concentration of chloride, fluoride, hydroxide, phosphate and sulfate in saltcake dissolutions. Chloride and hydroxide are very soluble components, so the ISM predicts them to be in the liquid phase almost 100% of the time. The experimental data and calculations report chloride and hydroxide in the solid phase. Sulfate is also fairly soluble and the ISM often predicts it to be completely in the liquid phase while the experimental data report some quantity in the solid phase. It is possible that the assumptions and calculations required to track the liquid and solid mass in MS Excel® contained errors that cause problems with the ISM predictions. This applies to all the components in the saltcake dissolutions, but is most applicable to very soluble components.

Fluoride and phosphate liquid concentrations were also over predicted by the ISM in saltcake dissolutions, but had less consistent predictions than hydroxide or sulfate. The fluoride and phosphate issues could be associated with the prediction of $Na_7F(PO_4)_2.19H_2O$, but sodium fluoride phosphate was predicted to be present by the ISM in two tanks that it was seen experimentally. The presence indicates qualitative, not quantitative agreement, so ISM predictions could still be inaccurate.

Implementation in HTWOS

The ISM has been implemented in HTWOS and its impact to mission metrics is described below by comparing the ISM scenario in HTWOS to the baseline scenario (without the ISM) in HTWOS. The change in mission metrics between the baseline scenario and the ISM scenario was driven by two competing phenomena; a decrease in the solubility of aluminum and an increase in the solubility of sulfate.

Table 3 compares several mission metrics from the baseline scenario with mission metrics from the ISM scenario. In the ISM scenario there was a 13% drop in the amount of HLW glass produced. This decrease was primarily due to an increase in the predicted solubility of sulfate and, to a lesser extent, an increase in the predicted solubility of phosphate, both HLW glass drivers. However, the decrease in HLW was offset by a 10% increase in the number of LAW containers produced. The larger LAW container count is mostly attributable to decreased aluminum solubility. Caustic, NaOH, is added in pretreatment to leach the solid phase aluminum (boehmite and gibbsite) to liquid phase aluminate [$Al(OH)_4^-$] and to keep aluminate in solution. The ISM scenario required the addition of 28% more caustic, to both leach solid phase aluminum and to maintain the aluminate in solution in WTP process vessels. The increased caustic results in more sodium, the primary LAW glass driver, delivered to the LAW melters.

Table 3. Mission metric differences between the baseline and ISM scenarios

	LAW Container Count	**HLW Canister Count**	**NaOH added [metric tons (MT)]**
Baseline	95,466	11,203	23,571
With ISM	104,571	9,776	30,060
Difference from Baseline Scenario	10%	-13%	28%

Table 4 shows the mass (on an oxide basis) that is incorporated into LAW and HLW glass for key components. Note that the implementation of the ISM results in an over 90% decrease in the amount of sulfur and phosphorous incorporated into HLW glass. The mass of aluminum oxide in both LAW and HLW is fairly consistent between scenarios, but the increased caustic addition required to keep that aluminum in solution means that 9.3% more sodium ends up in the LAW glass.

Table 4. Key component masses in LAW and HLW

Glass Oxide	**Waste Glass**	**Baseline (MT)**	**ISM (MT)**	**Difference from Baseline Scenario**
Al_2O_3	HLW	4.13×10^3	4.16×10^3	0.6%
	LAW	1.16×10^4	1.18×10^4	1.7%
Na_2O	HLW	1.45×10^3	9.14×10^2	-37%
	LAW	9.61×10^4	1.05×10^5	9.3%
P_2O_5	HLW	4.66×10^2	2.47×10^1	-95%
	LAW	3.16×10^3	3.59×10^3	14%
SO_3	HLW	9.12×10^1	5.49×10^0	-94%
	LAW	2.91×10^3	2.92×10^3	0.3%

Summary and Recommendations

From a global perspective the ISM did a good job fitting tank waste data. One goal of the evaluation was to highlight areas needing improvement, but it does not take away from the areas where the ISM did well. The ISM accurately predicted the phase distribution for Na, NO_3, and NO_2 in both study types and for C_2O_4, Cl, OH, SO_4, and CO_3 in one study type. The following recommendations are based on the areas where the ISM did not do well in the evaluation and on its implementation in HTWOS:

1. Further explore the aluminum solubility and gibbsite formation with a kinetic basis, due to the history of high aluminum solubility in tank waste. The aluminum predictions by the ISM are inconsistent and increase the amount of LAW containers in HTWOS due to increased caustic addition.

2. Review the sodium fluoride phosphate data set as fluoride and phosphate have inconsistent predictions by the ISM and phosphate solubility predicted by the ISM is significantly increased than the current HTWOS framework.
3. Obtain additional solubility data for oxalate and sodium oxalate to improve the inconsistencies associated with ISM sodium oxalate predictions.
4. Reevaluate chromium solubility, likely with a REDOX reaction that was not considered by the current model.
5. Review the data for sulfate solubility as the ISM predicts it as significantly more soluble than the current HTWOS framework.
6. Evaluate the ISM using tank retrieval data.
7. Evaluate of the ISM using data from the WTP.

References

1. RPP-RPT-51819, 2012, *Hanford Tank Waste Operations Simulator (HTWOS) Sensitivity Study*, Rev. 0, Washington River Protection Solutions, LLC, Richland WA.
2. RPP-RPT-47542, 2010, *Simplified Solubility Estimates for Selected Waste Compounds*, Rev. 0, Washington River Protection Solutions, LLC, Richland WA.
3. Pitzer, Kenneth S. (1973), "Thermodynamics of Electrolytes. I. Theoretical Basis and General Equations," The Journal of Physical Chemistry, Vol. 77, pp. 268-277.
4. Pitzer, Kenneth S. and Janice J. Kim (1974), "Thermodynamics of Electrolytes: IV. Activity and Osmotic Coefficients for Mixed Electrolytes," Journal of the American Chemical Society, Vol. 16, pp. 5701-5708.
5. RPP-RPT-50703, 2011, *Development of a Thermodynamic Model for the Hanford Tank Waste Operations Simulator (HTWOS)*, Rev. 0, Energy Solutions, LLC, Richland, WA.
6. RPP-PLAN-46002, 2010, *Wash and Leach Factor Work Plan*, Rev. 0, Washington River Protection Solutions, LLC, Richland WA.
7. RPP-51192, 2011, *Plan for the Evaluation of the HTWOS Integrated Solubility Model Predictions*, Rev. 0, Washington River Protection Solutions, LLC, Richland WA.
8. Reynolds, Jacob G. (2012), "The apparent solubility of aluminum (III) in Hanford high-level waste," Journal of Environmental Science and Health, Vol. 47, pp. 2213-2218.

(669e) Parametric Sensitivity Study of Interaction Properties for Simulation of Realistic Packed Bed Structures

Vaibhav Khane, Chemical Engineering, Missouri University of Science & Technology, Rolla, MO, Gary E. Mueller, Nuclear Engineering Department, Missouri University of Science & Technology, Rolla, MO and Al-Dahhan Muthanna, Chemical & Biological Engineering, Missouri University of Science and Technology, Rolla, MO

Experts in Discrete Element Modeling (EDEM™) is a discrete element method (DEM) based modeling software program, capable of providing information about granular solids flows occurring in many industrial applications. Packing of particles inside the container is the first and main step in EDEM™ based analysis. Packing algorithms available with commercial codes such as EDEM™ are used as a 'Black-Box' and are without any detailed validation exercise. In most cases, average porosity results are compared to benchmark the numerical packing results with available experimental results, which is not sufficient. The packing algorithm used in EDEM demands accurate input interaction properties which are not readily available in the literature for materials of interest. Hence, the objective of the current study is to perform parametric sensitivity study of interaction properties from reliable numerical analysis point of view. This sensitivity study will also highlight important interaction properties which need to be provided as an input to EDEM™ contact force model. The obtained results will provide a new and better understanding about the packing algorithm used in EDEM™ as well as recommendations to simulate realistic packed bed structures using EDEM™. The cylindrical packed bed has been chosen as a model for this parametric sensitivity study. Also, the slow and dense granular flow encountered in a Pebble Bed Reactor (PBR) can be approximated by static packed beds. The ongoing experimental work at Missouri S&T involves the implementation of a radioactive particle tracking (RPT) technique and a gamma ray densitometry (GRD) technique around a cold flow continuous pebble recirculation experimental set-up, which mimics operation of PBR, for the evaluation of solids dynamics and associated hold-up distribution. In addition, computed tomography experiments are being done on static packed beds to evaluate cross-sectional averaged solids and voids distribution. Experimental benchmark data obtained using above mentioned techniques can be used to validate EDEM™ based simulation results by mimicking same diameter aspect ratio (which is defined as the ratio of the diameter of container to the diameter of the particles) as that of experimental geometry. Hence, diameter aspect ratio of 23.9 (derived from experimental geometry dimensions) has been chosen for EDEM™ based analysis. EDEM™ based analysis requires the accurate input of coefficient of static and rolling friction and coefficient of restitution (COR). Various cases are designed and simulated in EDEM™ to test the effects of one particular parameter, while keeping other parameters constant at a value determined through the development of simple experimental set-ups. It is found that static friction for particle-wall and particle-particle interaction plays an important role in EDEM™ based simulations of packed beds. Interaction parameters such as the coefficient of rolling friction and COR are found to have minor or no influence on packed beds' structural properties. Results of this study also indicate that existing empirical correlations for mean porosity values should include static friction characteristics in addition to the diameter aspect ratio.

(669f) Aluminum Solubility: Impact of Sodium Nitrate, Sodium Nitrite, and Sodium Carbonate in the Presence of Sodium Hydroxide

Rebecca K. Toghiani[1,2], Laura T. Smith[2], Yunju Xia[2] and Jeffrey S. Lindner[2], (1)Dave C. Swalm School of Chemical Engineering, Mississippi State University, MS State, MS, (2)Institute for Clean Energy Technology, Mississippi State University, Starkville, MS

The stability of waste streams containing aluminate is a significant factor relating to unit operations (ion exchange) within the Waste Treatment and Immobilization Plant (WTP) under construction at the Hanford site. At an operational cost of $500M/yr for the WTP, a loss of one day of operation would amount to $1.37M. The overall recovery cost will be a function of the amount of gibbsite [$Al(OH)_3$] that forms. For small loadings where additional caustic leaching at elevated temperature will restore operation, the time delay may be five days. At significant precipitation where 50% or more of the interstitial volume within the ion exchange (IX) column is blocked, options will include re-insertion of new resin or extended leaching. A conservative estimate would then be from $6.6-41M/year. Thus, it is imperative that aluminum solubility be reliably predicted by available process simulation software.

The Environmental Simulation Program (ESP, OLI Systems, Inc.) is used for prediction of the electrolyte chemistry in these high ionic strength solutions. Since 1997, the Process Waste Chemistry team at the Institute for Clean Energy Technology (ICET, Mississippi State University) has undertaken to improve process waste chemistry predictions through measurement of the solubility of relevant systems under conditions of high ionic strength and at elevated temperatures. The efforts described herein were carried out under this program.

A test plan was written, designed, and reviewed in 2009 to identify the influence of sodium, hydroxide, nitrate, nitrite and carbonate on aluminum solubility as a function of temperature. Sodium hydroxide solutions were prepared over a range of 0.5 to 5.5 m. Gibbsite was used as the aluminum source. Two sets of samples were prepared at each nitrate, nitrite and carbonate loading. One set was allowed to equilibrate undisturbed. The other set was used to identify when equilibration had been achieved, through weekly sampling of the samples. Solubility experiments were conducted at three temperatures (25, 40, 55 °C), encompassing the expected range of operating temperature to be realized in the WTP.

The measured data in the form of molality for aluminum, hydroxide, carbonate, nitrate and nitrite were regressed using the Environmental Simulation Program (ESP V8.2) and the Mixed Solvent Electrolyte (MSE) model. Like-ion interactions not already present in the MSE Public (MSEPub) database were considered for data representation. Regression results will be presented along with the experimental data.

(738a) Removal of 14C From Irradiated Graphite for Waste Volume Reduction and Bulk Graphite Recycle: Thermal Treatment

Tara E. Smith and Mary Lou Dunzik-Gougar, Nuclear Engineering, Idaho State University: Idaho Falls Campus, Idaho Falls, ID

Removal of ^{14}C from Irradiated Graphite for Waste Volume Reduction and Bulk Graphite Recycle:

Thermal Treatment

Tara E. Smith and Mary Lou Dunzik-Gougar

Idaho State University: 1776 Science Center Dr., Idaho Falls, ID, 83402

smittar4@isu.edu mldg@isu.edu

INTRODUCTION

Nuclear power contributes 20% of the electricity used around the world today. To improve the efficiency of future nuclear power production, the U.S. and other countries are developing advanced nuclear systems coined "Generation IV concepts" as established by the Generation IV International Forum, which institutes protocol for individual countries to lead the development of a concept in which they have particular interest [1]. The U.S. Department of Energy commissioned the Idaho National Laboratory to develop the High Temperature Gas-Cooled Reactor (HTGR). The HTGR uses graphite coated fuel particles, graphite reflector and core structural components, and helium coolant [2]. As graphite is bombarded with neutrons, carbon-14 (^{14}C) is produced through three distinct reactions with ^{13}C, ^{14}N, and ^{17}O. In irradiated reactor graphite, the majority of ^{14}C has been found on the surface [3]. This occurrence indicates the most significant of the production reactions is neutron activation of ^{14}N, because ^{14}N is present on the surface and in surface pores as N_2 gas deposited from air. Carbon-14 is relatively long-lived (half life = 5730 years) and has significant mobility in groundwater and atmospheric systems. For these reasons, disposal of large irradiated graphite components from an HTGR would likely be costly. Further, the value of such pure nuclear–grade graphite is expected to increase and recycle may be an option. Thus, a means of removing the majority surface ^{14}C from the bulk graphite ^{12}C is being investigated. Pyrolysis and oxidation in a steam atmosphere have been suggested as ^{14}C decontamination methods [4]. Fachinger *et al.* demonstrated the concept of thermal treatment of irradiated nuclear graphite in the presence of steam or oxygen [5].

DESCRIPTION OF experimental WORK

Work discussed here is part of a larger project with the objective to determine the chemical nature of the ^{14}C in irradiated graphite and to use this information to determine an optimal method for ^{14}C removal. This paper summarizes thermal treatment development work.

Graphite samples are heated to a temperature in the range of 800-1500°C in the presence of inert argon gas, which carries any gaseous products released during treatment. In the presence of argon alone, graphite surface species will react with naturally adsorbed oxygen. To increase the oxidation rate of irradiated graphite surface species, small quantities of oxygen or carbon dioxide gas are added to the argon carrier gas. Oxygen combines with the surface to

*** Formatting Issue - Best Available Paper/Graphic ***

form carbon-oxygen bonds that are expected to result mostly in the formation of carbon monoxide (CO) [6].

Experimental parameters including treatment temperature, gas composition and gas flow rate will be optimized and are summarized in Table 1.

Table I. Key Thermal Treatment Experimental Parameters

Graphite	Gas	Temperature Regime
Unirradiated		Regime 1 (<600°C)
POCOFoam®		
NBG-18	Argon only	
NBG-25		Regime 2 (600-900°C)
Irradiated	Ar w/ O_2	
POCOFoam® (MURR)	Ar w/ CO_2	Regime 3 (>900°C)
NBG-18 (MURR)		
NBG-25 (ATR)		

Results

Work to date includes thermal treatment of un-irradiated POCOFoam® and nuclear grade NBG-18 graphite in the presence of argon doped with oxygen. During thermal treatment, the graphite is heated in a tube furnace, through which high purity argon and oxygen flows. After the gas leaves the thermal treatment furnace, it passes through a gas analyzer for chemical species identification. Before final collection, carbon species in the gas are fully oxidized to CO_2 via reaction with copper oxide at 800°C. CO_2 gas readily dissolves into sodium hydroxide (NaOH) solution, from which samples are taken periodically during the experiment. Solution samples from irradiated graphite treatment will be analyzed for carbon-14 via liquid scintillation counting. Figure 1 shows the experimental apparatus design.

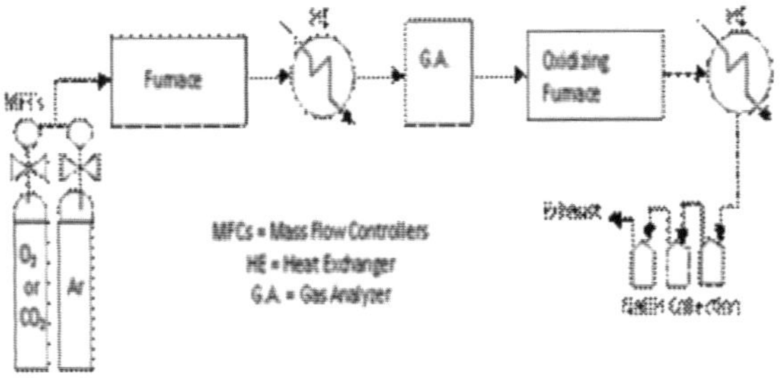

SHAPE * MERGEFORMAT

THE GRAPHIC ON THIS PAGE WAS NOT AVAILABLE ON THE E-MEDIA VERSION OF THIS PAPER.

Fig.1. Experimental apparatus for thermal treatment of nuclear graphite to remove ^{14}C

*** Formatting Issue - Best Available Paper/Graphic ***

Experiments were performed at 700°C and 1400°C for two levels of dopant, 1.5 sccm and 2.6 sccm of O_2, to evaluate the effects of temperature and quantity of available oxidizing agent (and associated oxidation mechanism) on the oxidation of un-irradiated graphite. Thermal treatment of irradiated graphite is in progress. The normalized selective removal rate of ^{14}C will be evaluated at the same conditions described above. The most recent results from the experiments will be presented.

NOMENCLATURE

HTGR = High Temperature Gas Reactor

ATR = Advanced Test Reactor at Idaho National Lab

CO = Carbon Monoxide

CO_2 = Carbon Dioxide

NaOH = Sodium Hydroxide

MURR = University of Missouri Research Reactor

NBG = Designation for nuclear grade graphites produced by SGL Group

REFERENCES

1. NUCLEAR ENERGY AGENCY. "*Generation IV International Forum.*" Technical Secretariat. 2009

2. Gen-IV International Forum. "*The Very-High-Temperature Reactor (VHTR) is a graphite-moderated, helium-cooled reactor with a thermal neutron spectrum.*" http://www.gen-4.org/Technology/systems/vhtr.htm

3. ELECTRIC POWER RESEARCH INSTITUTE, "*Graphite Decommissioning: Options for Graphite Treatment, Recycling, or Disposal, including a Discussion of Safety-Related Issues,*" EPRI Technical Report 1013091(March 2006)

4. J. FACHINGER, L.V.WERNER, "*Decontamination of Nuclear Graphite,*" 3rd International Topical Meeting on High Temperature Reactor Technology. (October 2006)

5. FACHINGER, J., PODRUHZINA, T., VON LENSA, W., "Decontamination of Nuclear Graphite by Thermal Treatment," Solutions for Graphite Waste, Manchester, UK (March 2007)

6. U.S. Department of Energy Environmental Management Spent Fuel Management Office. "Graphite Oxidation Thermodynamics/Reactions," DOE/SNF/REP-018, DOE (1998)

(738b) Removal of Radioactive Iodine Species by Solid Sorbents

Ronghong Lin and Lawrence L. Tavlarides, Department of Biomedical and Chemical Engineering, Syracuse University, Syracuse, NY

Iodine-129 is one of several radioactive iodine isotopes formed in fission reactions and it has a very long half-life of $16{\times}10^6$ years. A variety of iodine-129 species are present in off-gases from spent nuclear fuel reprocessing facilities including I_2, CH3I, HI, higher molecular weight alkyl iodides, and others. Removal and immobilization of these species is one of a number of critical barriers to nuclear waste treatment. The current study will develop predictive models for simulating dynamic adsorption processes, and the results will be used to provide recommendations on a path forward for the development of sorbents and sorption processes for off-gas treatment. In this presentation, we will discuss the development of a continuous flow adsorption system which is being applied to study single-pellet adsorption dynamics. Kinetics of I_2 adsorption by silver-exchanged mordenite will be discussed. Also, experimental studies on adsorption of various iodine species (I_2, CH_3I, C_2H_5I, etc.) by a number of sorbents (silver-exchanged mordenite, ZIF-8, etc.) will be discussed, and preliminary adsorption equilibrium data will be presented.

(738c) Development and Demonstration of A Prototype Microfluidic-Based Robotic Sampling System

Jack D. Law, John Svoboda, Julia L. Tripp, Veronica Rutledge and Troy G. Garn, Idaho National Laboratory, Idaho Falls, ID

A state-of-the-art microfluidic-based robotic sampling system has been designed, a prototype manufactured and testing performed. This sampling system uses microfluidic sample chips and will allow for near real time analysis of samples in-cell to support used nuclear fuel processing or nuclear waste treatment. The automated robotic system picks up the sample chip, transports it to the sample chamber, takes the sample, transports the chip to a manifold for injection of the microsamples into a UV Visible spectrophotometer, removes the sample chip and disposes of the sample chip. In addition to providing near real time analysis of samples, this will reduce waste volumes generated significantly as compared to traditional sampling methods which typically require several milliliter sample volumes and large dilutions to transport the samples out of cell for analysis. Additionally, the system will reduce personnel exposure and could provide a convenient method for sample archival. The prototype robotic sampling system will be described and results of integrated testing of the system will be presented.

(738d) Real-Time Actinide Monitoring with Tensioned Metastable Fluid Detectors for Enhanced Safeguards and Security in Spent Nuclear Fuel Chemical Reprocessing Facilities

Jeffrey A. Webster[1], Joe Lapinskas[2], Brian Archambault[1], Tom Grimes[1], Alex Bakken[1] and Rusi P. Taleyarkhan[1], (1)Nuclear Engineering, Purdue University, West Lafayette, IN, (2)QSA Global, Inc., Burlington, MA

Introduction

It has previously been demonstrated that tensioned metastable fluid detectors (TMFDs) offer a variety of unique mechanisms for detecting alpha and neutron radiation while remaining selectively insensitive to gamma photons and beta decay [1,2]. This paper will discuss some of the capabilities of TMFD systems and how they can be utilized for enhancing safeguards and security in spent nuclear fuel (SNF) chemical reprocessing facilities.

The TMFD technology is based on the principle of radiation seeded cavity nucleation and growth to audible-visible levels in tensioned metastable state liquids. When a nuclear particle interaction occurs in a sufficiently tensioned liquid, rapid vaporization results in a visible bubble and an audible mechanical shock [3]. The phenomenon is similar to bubble chambers and superheated droplet detectors but in this case the metastability in the liquid is created with tension (sub-vacuum pressures at room temperature) instead of thermal superheat.

Two particular TMFD device configurations have been developed for application to a variety at radiation detection scenarios. The first is the centrifugally tensioned metastable fluid detector (CTMFD), which uses the centrifugal force principle - created in a specialized rotating enclosure to create the tension needed to detect nuclear particles. By suitably tailoring the liquid's metastable state, the as-desired sensitivity to energy and signature of radiation particle can be controlled. The second device is the acoustically tensioned metastable fluid detector (ATMFD), which uses resonance-mode focused acoustic waves in the liquid (much like in a laser cavity) to induce oscillatory compression-tension states. Like for the CTMFD, the ATMFD also offers the capability for energy and particle sensitivity selection based by tailoring of the micro-second timed spatially-variant acoustic wave fields [4].

Applications of TMFD Technology to Spent Nuclear Fuel Chemical Reprocessing Plants

The application of TMFDs for providing real-time monitoring of neutron and alpha radiation emitting special nuclear materials (SNMs) like U, Pu, Np and Cm is discussed in sub-sections that follow:

<u>SNM alpha spectroscopy</u>

The TMFD systems have been found capable of radiation energy discrimination capabilities. Unlike many traditional radiation detectors which rely on electrical signal height or pulse shape discrimination, the TMFD uses the inherently unique mechanics of its detection technique to offer transformational spectroscopic capability. Various isotopes of SNMs such as Pu-239, Pu-238, Am-241, and Cm-242 emit alpha radiation at distinct energies, the detection of which provides for the tell-tale sign for their presence (or diversion). The detection mechanism in the TMFD is threshold-based meaning under chosen conditions particles depositing at or above the selected energy are detected while others are not – while also remaining unfettered by the presence of "nuisance" radiation such as beta-gamma emission. Straightforwardly, the amount of mechanical tension or negative pressure (Pneg) that a liquid is placed will determine the specific energy of alpha radiation particles it is sensitive to.

Detection of alpha decay from actinides and spectroscopy has been systematically demonstrated in the CTMFD by dissolving known quantities of specific actinides into the detector fluid and recording the detector response with Pneg.

Calibration has been done with several NIST certified single isotope transuranic actinide solutions [5]. After doing these calibrations, it was not only found that the difference in energy between the actinide elements was easily detectable but also that the difference in energy between individual isotopes of the same element could be detected.

Several applications of CTMFD based alpha monitoring and spectroscopy in a UREX or PUREX SNF chemical reprocessing facility are under study using a build-in protocol. First, since Uranium exhibits the lowest energy alphas of the actinides in the UREX process, it is easily ignored in scenarios where there is no interest in the Uranium concentration. Also, the ability to ignore uranium alphas could allow for detection of unwanted transuranic extraction during the UREX phase.

Since the Cm isotopes emit the highest energy (~6MeV) alphas amongst the actinides of interest, they are the first to detect. By setting the CTMFD to be sensitive to only Cm, the unwanted extraction of Cm prior to the TRUEX process can be monitored in real-time. Very small quantities of actinides can be detected in the CTMFD. The CTMFD has been shown to be sensitive to quantities below 1 part per million for Uranium-238 and below 10 parts per quadrillion for Pu-238.

Another key application of CTMFD based alpha spectroscopy is for Pu isotopes monitoring. Since the alpha energies of Pu are higher than that of Np, the quantity of Pu in the NPEX product stream can be determined. Plutonium would also be detectable in the UREX and CCD-PEG products if it were present therein, unexpectedly. Some experimentation has also been done to measure the ratio of Pu-238 to Pu-239. Detection of mixtures of Pu-238:Pu-239 in ratios of 1:1, 2:1, and 3:1 have been successfully conducted.

Active Detection of Fissionable/Fissile Material

In many scenarios, it may be desirable to monitor the location of fissionable/fissile materials in a reprocessing facility without removing samples. Active interrogation offers non-destructive assay if the interrogating radiation can be discriminated. The ATMFD's energy discrimination capabilities allow for lower energy particles from an active interrogation source to be ignored while detecting higher energy fission neutrons if fissionable material is present. Simulations have been conducted using the Monte Carlo nuclear particle transport code MCNP-PoliMi to predict the utility of the ATMFD in active detection of fissionable material in a reprocessing facility.

The ATMFD has been modeled in MCNP-PoliMi along with a representative 5cm diameter pipe filed with the UREX feed liquid from a PWR with 3% enriched fuel with 33GW-day/MTU burnup [6]. The scenario was modeled as one of many example applications of this technique. Two active interrogation sources were considered. The first was a 2.45 MeV D-D fusion neutron source and the other a 60 keV neutron source [7]. For active interrogation to work, the ATMFD must be set to avoid detection of the source particles. This is accomplished by lowering the acoustic drive power so that the detector is not sensitive to the source neutrons.

The simulations offer insight that using a modest intensity interrogating neutron source of, ~ 10^8 neutrons per second, that fissionable material in a reprocessing facility should be detectable [6]. For example, if the interrogating neutron source is a D-D accelerator emitting 2.45 MeV neutrons, the TMFD will be sensitive only to fission-induced neutrons above 2.45 MeV. The same scenario utilizing a 60 keV source would allow for detection of only fissile (vs fissionable) material which could then be used to monitor the movement of potential weapons grade SNMs with specificity, while ignoring many of the other actinides which may not be of interest.

Directional Neutron Monitoring

Past research work by our group has shown that the ATMFD is not only capable of detecting the presence of neutrons while remaining completely insensitive to intense gamma fields (>10^{23} g/cc/sec) [2] but also be capable of detecting directionality of incident neutrons from SNMs [1]. Directional neutron detectors offer vastly superior background suppression when compared to traditional neutron detectors (e.g. He^3, BF_3), and the ability to image the neutron source directly. An ability which would be particularly advantageous in both passive interrogation scenarios where one needs to discriminate neutrons originating from a single targeted location and active interrogation scenarios where one needs to discriminate interrogating neutrons (or photons) from neutrons resulting from the fission of the targeted material. Assessments have been conducted utilizing MCNP-PoliMi to investigate the possibility of employing the direction-position sensing capabilities of the ATMFD system to indicate the presence and location of neutron emitting isotopes in various stages of SNF fuel chemical reprocessing plant.

The full paper and presentation will provide details of theoretical and experimental assessments.

References

1. Archambault, B. A., J. A. Webster, J. R. Lapinskas, T. F. Grimes, R. P. Taleyarkhan and A. Eghlima, "Transformational Nuclear Sensors - Real-Time Monitoring of WMDs, Risk Assessment & Response," *IEEE 978-1-4244-6056-5/10, pp.421-427, 2010.*

2. Sansone, A., S. Zielinski, J. A. Webster, J. Lapinskas, R. P. Taleyarkhan and R. C. Block, "Gamma-Blind Nuclear Particle-Induced Bubble Formation in Tensioned Metastable Fluids," *Transactions of the American Nuclear Society*, Vol. 104, pp. 1033, Hollywood, Florida, June 26-30, 2011.

3. Taleyarkhan, R. P., C. D. West, and J. Cho, "Energetics of Nano-to-Macro Scale Triggered Metastable Fluids," *Oak Ridge National Laboratory Report* ORNL/TM-2022/233, 2022.

4. Lapinskas, J., et al., "Towards leap-ahead advances in radiation detection," *Proceedings of the 16th International Conference on Nuclear Engineering (ICONE-16),* Orlando, Florida, 2008.

5. Lapinskas, J., S. Zielinski, J. A. Webster, R. P. Taleyarkhan, S. McDeavitt, Y. Xu, "Tension metastable fluid detection systems for special nuclear material detection and monitoring," *Nuclear Engineering and Design,* 240, pp. 2866-2871, 2010.

6. Webster, J. A., R. P. Taleyarkhan, "Tensioned Metastable Fluid Detectors in Nuclear Security for Active Interrogation of Special Nuclear Materials - Part B," World Journal of Nuclear Science and Technology, Vol. I, pp. 66-76, 2011.

7. Hagmann, C. A., et al., "Active Detection of Shielded SNM With 60-keV Neutrons," *IEEE Transactions on Nuclear Science.* Vol. 56, pp. 1215-1217, 2009.

FORMIC ACID FREE FLOWSHEET DEVELOPMENT TO ELIMINATE CATALYTIC HYDROGEN GENERATION IN THE DEFENSE WASTE PROCESSING FACILITY

Dan P. Lambert, Michael E Stone, and J. David Newell, Savannah River National Laboratory
Terri L. Fellinger and Jonathan M. Bricker, Savannah River Remediation, LLC
Aiken, SC, USA

Abstract

The Defense Waste Processing Facility (DWPF) processes legacy nuclear waste generated at the Savannah River Site (SRS) during production of plutonium and tritium demanded by the Cold War. The nuclear waste is first treated via a complex sequence of controlled chemical reactions and then vitrified into a borosilicate glass form and poured into stainless steel canisters. Converting the nuclear waste into borosilicate glass canisters is a safe, effective way to reduce the volume of the waste and stabilize the radionuclides.

Testing was initiated to determine whether the elimination of formic acid from the DWPF's chemical processing flowsheet would eliminate catalytic hydrogen generation. Historically, hydrogen is generated in chemical processing of alkaline High Level Waste sludge in DWPF. In current processing, sludge is combined with nitric and formic acid to neutralize the waste, reduce mercury and manganese, destroy nitrite, and modify (thin) the slurry rheology. The noble metal catalyzed formic acid decomposition produces hydrogen and carbon dioxide. Elimination of formic acid by replacement with glycolic acid has the potential to eliminate the production of catalytic hydrogen.

Flowsheet testing was performed to develop the nitric-glycolic acid flowsheet as an alternative to the nitric-formic flowsheet currently being processed at the DWPF. This new flowsheet has shown that mercury can be reduced and removed by steam stripping in DWPF with no catalytic hydrogen generation. All processing objectives were also met, including greatly reducing the Slurry Mix Evaporator (SME) product yield stress as compared to the baseline nitric/formic flowsheet. Ten DWPF tests were performed with nonradioactive simulants designed to cover a broad compositional range. No hydrogen was generated in testing without formic acid.

Introduction

Savannah River Remediation (SRR) is evaluating changes to its current DWPF flowsheet to improve processing cycle times. This will enable the facility to support higher canister production while maximizing waste loading. Higher throughput is needed in the CPC since the installation of the bubblers into the melter has increased melt rate. Due to the significant maintenance required for the DWPF gas chromatographs (GC) and the potential for production of flammable quantities of hydrogen, reducing or eliminating the amount of formic acid used in the CPC is being developed. Earlier work at Savannah River National Laboratory has shown that replacing formic acid with an 80:20 molar blend of glycolic and formic acids has the potential to remove mercury in the SRAT without any significant catalytic hydrogen generation.[1,2,3] This report summarizes the research completed to determine the feasibility of processing without formic acid.

In earlier development of the glycolic-formic acid flowsheet, one run (GF8) 2 was completed without formic acid. It is of particular interest that mercury was successfully removed in GF8, no formic acid at 125% stoichiometry. Glycolic acid did not show the ability to reduce mercury to elemental mercury in initial screening studies, which is why previous testing focused on using the formic/glycolic blend.

The objective of the testing detailed in this document is to determine the viability of the nitric-glycolic acid flowsheet in processing sludge over a wide compositional range as requested by DWPF.[4] This work was performed under the guidance of Task Technical and Quality Assurance Plan (TT&QAP).[5] The details regarding the simulant preparation and analysis have been documented previously.[6]

Experimental Procedure

The experimental apparatus used in these experiments is typical for DWPF SRAT/SME testing. The four experiments were performed in 4-L kettles. The test equipment included a GC to measure off-gas composition, an ammonia scrubber, and a pH meter. In all runs, the SRNL acid calculation spreadsheet[7] used the Koopman equation[8] to determine acid addition quantities and dewater targets.

---

[1] Pickenheim, B.R., M.E. Stone, SRAT Alternative Reductant Feasibility Assessment – Phase I, SRNL-STI-2009-00120, Savannah River National Laboratory, Aiken, SC, February 2009.

[2] Pickenheim, B.R., M.E. Stone, J.D. Newell, Glycolic-Formic Acid Flowsheet Development, SRNL-STI-2010-00523, Rev 0, Savannah River National Laboratory, Aiken, SC, November 2010.

[3] D.P. Lambert, B.R. Pickenheim, M.E. Stone, J.D. Newell, D.R. Best, Glycolic - Formic Acid Flowsheet Final Report for Downselection Decision, SRNL-STI- 2010-00523, Rev 1, Savannah River National Laboratory, Aiken, SC, March 2011.

[4] Fellinger, T.L., "*Alternate Reductant Flowsheet Development – Phase I*", HLW-DWPF-TTR-2012-0003, Revision 0, October 2011.

[5] Lambert, D.P., Task Technical and Quality Assurance Plan for Glycolic Acid Flowsheet Development, SRNL-RP-2011-01586, Savannah River National Laboratory, Aiken, SC, November 2011.

[6] D. C. Koopman, D. P. Lambert, Initial Characterizations and SRAT Simulations of Four Sludge Matrix Study Simulants, SRNL-STI-2009-00606, Revision 0, Savannah River National Laboratory, Aiken, SC, December 2009

[7] Lambert, D.P., Acid Calculation Spreadsheet for DWPF Simulations, Revision 1, SRNL-PSE-2006-00176, Savannah River Site, Aiken, SC 29808 (2006).

[8] Koopman, D.C., A.I. Fernandez, B.R. Pickenheim, Preliminary Evaluations of Two Proposed Stoichiometric Acid Equations, Revision 0, Savannah River Site, Aiken, SC 29808 (2009).

CPC Simulation Details

The SRAT 4-L rigs were assembled following the guidelines of SRNL-3100-2011-00127.[1] The intent of the equipment is to functionally replicate the DWPF processing vessels. Each glass kettle is used to replicate both the SRAT and SME, and it is connected to the SRAT Condenser, the MWWT, and the FAVC. The Slurry Mix Evaporator Condensate Tank (SMECT) is represented by a sampling bottle that is used to remove condensate through the MWWT. For the purposes of this paper, the condensers and wash tank are referred to as the off-gas components. A sketch of the experimental setup is given in Figure 2-1.

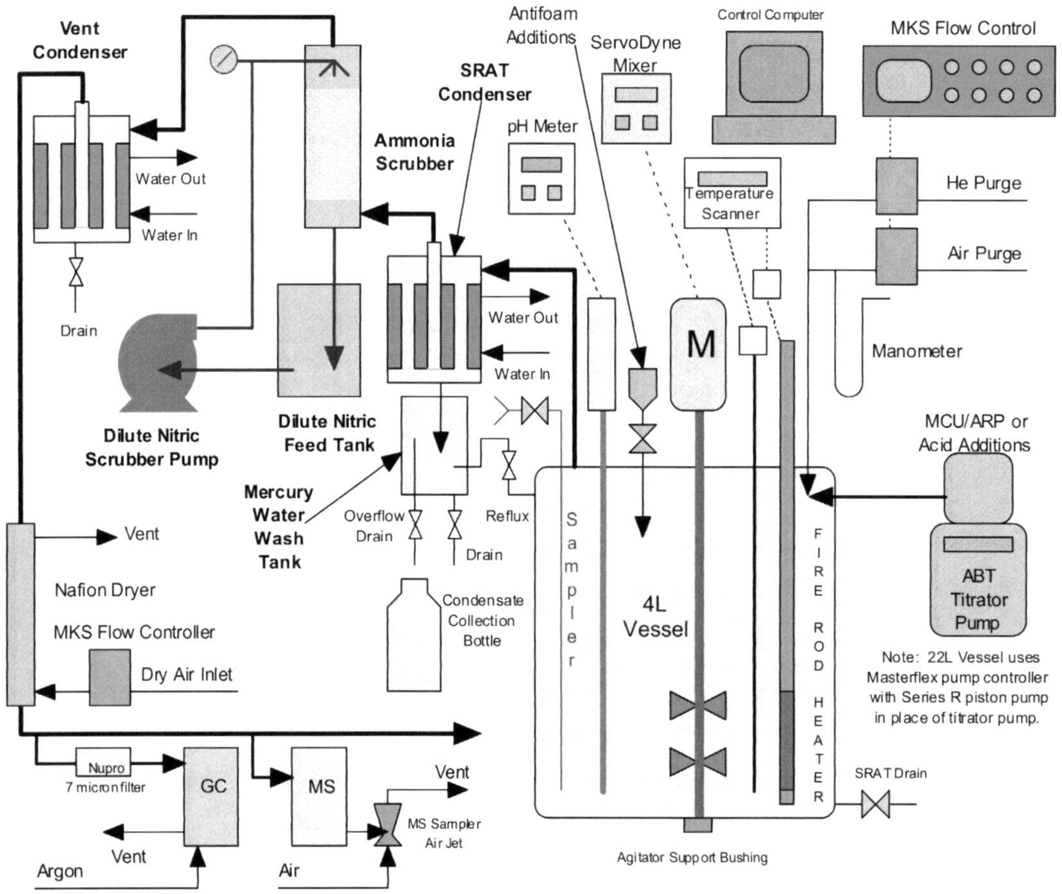

Figure Error! No text of specified style in document.**-1. Schematic of CPC Equipment Set-Up**

Concentrated nitric acid (~50 wt %) and glycolic acid (~70 wt %) were used to acidify the sludge and perform neutralization and reduction reactions during processing. The total amount of acid (in moles) to add for each run was determined using the Koopman acid equation[8]. The Koopman minimum acid equation was used with a 100% stoichiometric factor for all tests except GF38 (125%), GF40 (134%) and GF41 (130%).

---

[1] Stone, M. E., Lab-Scale CPC Equipment Set-up, SRNL-ITS-2006-000742011-00127, Savannah River Site, Aiken, SC 29808 (2011).

The acid mix was partitioned between nitric and glycolic acid by utilizing the latest REDOX equation[1] with a term added for glycolate ion (see below). A coefficient of 6 was used on the glycolate term based on electron equivalence.[2] The REDOX target ($Fe^{2+}/\Sigma Fe$) was 0.1. Process assumptions were made to predict SME product anion concentrations. In addition to the standard assumptions needed for formate and oxalate loss and nitrite to nitrate conversion, a factor was added to the acid calculation for glycolate loss. Process assumptions for the stoichiometric window testing were adjusted based on results from earlier testing.

$$REDOX=0.2358+0.1999*((2*C_{formate}+4*C_{oxalate}+4*C_{Carbon}+6*C_{glycolate}-5*(C_{Nitrate}+C_{Nitrite})-5*C_{Mn}))*(45/TS)$$

Where C_x = species concentration of component x, g-mole/kg melter feed, TS = total solids in melter feed in wt %, and REDOX is a molar ratio of $Fe^{2+}/\Sigma Fe$

A standard 4-L SRAT/SME apparatus with an ammonia scrubber was used for these simulations. The scrubber solution consisted of 749 g of de-ionized water and 1 g of 50 wt% nitric acid. The solution was recirculated through the column by a MasterFlex pump at 300 mL/min through a spray nozzle at the top of the packed section. Glass rings were used as packing and did not significantly add to the back pressure on the SRAT vessel as has been seen in earlier tests with different packing. The SRAT condenser was maintained at 25 °C during the run, while the vent condenser was maintained at 4 °C.

Sludge Preparation

SRNL produced four matrix sludge simulants in order to improve the understanding of how changing sludge composition impacts DWPF waste processing. These simulants have been used in other SRNL studies, and the composition has been previously measured.[3] These simulants were used to demonstrate the flowsheet across a broad compositional range. In addition, two less washed simulants (1.6 and 1.9 M Na) were produced to study the impact of less washing on CPC processing.

---

1 Jantzen, C. M. and M. E. Stone, Role of Manganese Reduction/Oxidation (REDOX) on Foaming and Melt Rate in High Level Waste Melters, WSRC-STI-2006-00066, Savannah River Site, Aiken, SC, 29808 , March 2007.

2 Jantzen, C.M., J.R. Zamecnik, D.C. Koopman, C.C. Herman, and J.B. Pickett, Electron Equivalents Model for Controlling Reduction-Oxidation (REDOX) Equilibrium during High Level Waste (HLW) Vitrification, WSRC-TR-2003-00126, Savannah River Site, Aiken, SC 29808 (2003).

3 Lambert, D. P., Koopman, D. C., Glycolic-Formic Acid Flowsheet Sludge Matrix Study, SRNL-STI-2011-00275, Savannah River Site, Aiken, SC, 29808 (June 2011).

Table Error! No text of specified style in document.**-1. Composition of Sludge Simulants**

Result	GF34 HiFeHiMn	GF35 SB7An	GF36 HiFeLoMn	GF37/38 LoFeLoMn	GF40 1.6 M Na	GF41 1.9 M Na	Units
Total Solids	23.70	18.02	22.81	23.07	24.14	25.43	wt%
Calcined Solids	17.81	13.61	16.95	16.00	17.01	17.85	wt%
Insoluble Solids	16.70	12.57	16.35	16.05	16.51	16.97	wt%
Soluble Solids	7.00	5.45	6.47	7.01	7.63	8.46	wt%
Slurry Density	1.185	1.142	1.189	1.176	1.174	1.215	kg / L slurry
Filtrate Density	1.057	1.053	1.055	1.057	1.076	1.091	kg / L supernate
Aluminum	9.000	15.65	9.130	23.8	14.8	13.9	wt % calcined basis
Boron	<0.100	<0.100	<0.100	<0.100	NM	NM	wt % calcined basis
Barium	0.077	0.102	0.101	0.0705	0.085	0.0802	wt % calcined basis
Calcium	3.83	0.836	2.22	1.97	0.565	0.288	wt % calcined basis
Cadmium	<0.010	NM	<0.010	<0.010	NM	NM	wt % calcined basis
Cerium	0.104	0.148	0.108	0.0965	NM	NM	wt % calcined basis
Chromium	0.015	0.0455	0.285	0.244	0.027	0.0260	wt % calcined basis
Copper	0.045	0.033	0.045	0.048	0.040	0.0350	wt % calcined basis
Iron	32.4	19.2	31.5	12.2	14.8	13.7	wt % calcined basis
Potassium	0.120	0.125	0.0905	0.0955	0.369	0.392	wt % calcined basis
Magnesium	0.396	0.366	2.69	2.42	0.317	0.302	wt % calcined basis
Manganese	4.04	4.37	0.721	0.661	4.86	4.53	wt % calcined basis
Sodium	12.9	15.3	13.1	14.2	22.8	24.4	wt % calcined basis
Nickel	0.213	3.37	2.6345	2.31	2.10	1.95	wt % calcined basis
Phosphorus	<0.100	<0.100	<0.100	<0.100	0.032	<0.010	wt % calcined basis
Lead	0.071	0.025	0.047	0.0715	<0.010	<0.010	wt % calcined basis
Sulfur	0.289	0.371	0.340	0.374	0.276	0.333	wt % calcined basis
Silicon	1.580	1.91	1.52	1.32	1.52	1.369	wt % calcined basis
Tin	<0.010	0.013	0.106	0.0925	NM	NM	wt % calcined basis
Titanium	<0.010	0.025	<0.010	<0.010	0.025	0.0230	wt % calcined basis
Zinc	0.065	0.047	0.0775	0.0705	0.049	0.0452	wt % calcined basis
Zirconium	0.054	0.252	0.1175	0.049	0.027	0.195	wt % calcined basis
Nitrite	17,900	9,140	17,800	13,300	13,500	15,800	mg/kg slurry
Nitrate	13,550	6,470	13,400	13,300	7,895	9,940	mg/kg slurry
Formate	<100	<100	<100	<100	<100	<100	mg/kg slurry
Sulfate	1,770	1,460	1,575	1,590	1,980	2,610	mg/kg slurry
Chlorine	116	<100	131	127	<100	<100	mg/kg slurry
Phosphate	0	<100	<100	<100	<100	<100	mg/kg slurry
Oxalate	300	8,500	275	295	18,750	20,000	mg/kg slurry
Glycolate	<100	NM	<100	<100	<100	<100	mg/kg slurry
Slurry TIC	2,751	1,066	2,492	2,400	1,840	1,730	mg/kg slurry
Supernate TIC	1,080	664	1,310	1,280	1,790	1760	mg/L supernate
Total Base pH 7	0.590	0.580	0.562	0.522	0.838	0.879	moles/L

Table Error! No text of specified style in document.**-2. Mercury and Noble Metal Composition Added to Sludge Simulants, wt% Total Solids Basis**

Noble Metal	Runs GF34-GF38	GF40-41
Target Hg	1.5000	1.500
Target Ag	0.0014	0.0144
Target Pd	0.0790	0.0033
Target Rh	0.0380	0.0192
Target Ru	0.2170	0.0877

An additional supernate simulant was prepared to supplement the four slurry simulants above. The purpose of this simpler simulant was to improve understanding of the mercury reduction chemistry. The simulant was similar to the supernate used in the matrix slurry preparation. The only soluble species added were sodium hydroxide, sodium nitrite, sodium nitrate, sodium sulfate, sodium oxalate, sodium carbonate and potassium nitrate. The resulting concentration is summarized in **Table** Error! No text of specified style in document.**-3**. The added noble metal and mercury target of these runs is summarized in **Table** Error! No text of specified style in document.**-4**. Note that because of the lower total solids of the supernate, the added mass of noble metals and mercury is approximately one-third that added in the slurry experiments.

Table Error! No text of specified style in document.**-3. Composition of Supernate Simulant**

Anion or Cation	GF39a-d	GF39e
Nitrite, mg/kg	21,561	0
Nitrate, mg/kg	15,784	16,311
Carbonate, mg/kg	6,051	6,253
Oxalate, mg/kg	351	363
Sulfate, mg/kg	1,888	1,951
Free Hydroxide, M	3,556 (0.221 M)	0.221
Na, mg/kg	27,067	27,067
K, mg/kg	153	153

Table Error! No text of specified style in document.**-4. Mercury and Noble Metal Composition Added to Supernate Simulants, wt% Total Solids Basis**

Noble Metal	GF39a	GF39b	GF39c	GF39d	GF39e
Target Hg	1.5000	1.5000	1.5000	1.5000	2.5717
Target Ag	0.0000	0.0014	0.0014	0.0014	0.0000
Target Pd	0.0000	0.0790	0.0790	0.0790	0.0000
Target Rh	0.0000	0.0297	0.0297	0.0297	0.0000
Target Ru	0.0000	0.2170	0.0000	0.2170	0.0000

CPC Run Details

The twelve nitric-glycolic acid flowsheet tests with slurry and five tests with supernate were performed at the ACTL using the four-liter kettle setup. **Table** Error! No text of specified style in document.**-5** identifies each run and its corresponding assumptions.

Table Error! No text of specified style in document.**-5. CPC Simulation Process Assumptions**

Run	Sludge	Cycles	Date	% Koopman Acid Stoichiometry
GF34	HiFeHiMn	SRAT/SME	16-Nov-11	104.0
GF34b	HiFeHiMn	SRAT/SME	16-Nov-11	104.0
GF34c	HiFeHiMn	SRAT/SME	16-Nov-11	104.0
GF35	SB7A	SRAT/SME	17-Nov-11	100.0
GF36	HiFeLoMn	SRAT/SME	16-Nov-11	106.1
GF36b	HiFeLoMn	SRAT	25-Jan-12	106.1
GF36c	HiFeLoMn	SRAT	25-Jan-12	106.1
GF37	LoFeLoMn	SRAT/SME	2-Feb-12	100.0
GF37b	LoFeLoMn	SRAT	17-Nov-11	100.0
GF38	LoFeLoMn	SRAT	2-Feb-12	125.0
GF39A	Supernate	SRAT	22-Feb-12	100.0
GF39B	Supernate	SRAT	22-Feb-12	100.0
GF39C	Supernate	SRAT	29-Feb-12	100.0
GF39D	Supernate	SRAT	29-Feb-12	80.0
GF39E	No Nitrite Supernate	SRAT	8-May-12	100.0
GF40	1.6M Na	SRAT	24-May-12	133.9
GF41	1.9 M Na	SRAT	24-May-12	130.0

DWPF design basis processing conditions were scaled down and used for most processing parameters including SRAT/SME air purges and boil-up rate. SRAT product total dried solids were targeted at 27 wt% for the slurry simulant runs. Final SME total dried solids were targeted at 45% at 36% waste loading.

Because nitric and glycolic acid are more dilute acids than formic acid, both acids were added at the same molar flowrate as formic acid. Thus nitric acid was added at a DWPF scaled flowrate of 4.572 gallons per minute and glycolic acid was added at a DWPF scaled flowrate of 3.948 gallons per minute to maintain acid addition times. It is recommended that DWPF modify the acid feed pumps to deliver the higher flow rates before implementing the glycolic flowsheet.

The following constraints must be met by the current DWPF CPC flowsheet:

- SRAT hydrogen <0.65 lb/hr
- SME hydrogen <0.223 lb/hr
- Reduce mercury to elemental form
- Steam strip mercury below 0.8 wt% in the SRAT product dried solids
- SRAT product less than 1000 mg nitrite/kg product slurry

- SRAT product rheology[1] design basis 1.5 to 5 Pa yield stress and 5 to 12 cP consistency
- SME product rheology[1] 2.5 to 15 Pa yield stress and 10 to 40 cP consistency
- Glass REDOX of 0.09-0.33 $Fe^{2+}/\Sigma Fe$
- Minimize water in SME product (55 wt% typical)
- Minimal foaming

Twelve to fifteen samples were taken during each SRAT cycle to monitor the progress of the main chemical reactions. Major cations and anions were checked immediately after acid addition. Samples were taken during boiling to monitor suspended and dissolved mercury in the SRAT slurry. These samples were transferred directly into digestion vials to eliminate potential segregation of mercury during sub-sampling/aliquoting steps. The SRAT and SME product slurries were sampled similarly once they had cooled to 90° C while the vessel contents were still mixing.

Additional SRAT product samples were taken for compositional and solids analyses after the product had cooled further. The MWWT and FAVC were drained and the condensates weighed after both the SRAT and SME cycles. Elemental mercury was separated from the aqueous phase in the post-SRAT MWWT sample, and the mass of the mercury-rich material determined. Beads of elemental mercury were also recovered from a few of the SME dewatering condensates and weighed (depending on how big or numerous the bead(s) appeared to be).

Data are presented in Section 3 showing how the nitric-glycolic flowsheet met or exceeded the processing constraints in the list above with the possible exception of mercury removal and REDOX.

Results and Discussion

Four SRAT simulations with supernate and eight SRAT/SME process simulations with slurry feeds were completed to demonstrate the feasibility of using only glycolic acid as the reducing acid in SRAT processing. The elimination of formic acid has the potential to eliminate the catalytic generation of hydrogen, which could lead to the reduction of the air purge in the DWPF CPC. The main concern in eliminating formic acid[1] is that the mercury won't be effectively reduced, and won't be removed by steam stripping to meet the DWPF SRAT mercury target and minimize the mercury sent to the melter. The discussion begins with the supernate results followed by the slurry results.

Supernate Testing

Four SRAT process simulations were completed with a simple supernate solution with added mercury and noble metals. A fifth run was completed with a nitrite free supernate solution to determine whether nitrite is needed to reduce mercury. These runs were performed after the slurry runs in order to better understand the processing chemistry. In particular, it was important to understand when the mercury is reduced in processing. Samples were pulled during glycolic acid addition and for several hours during the dewater and reflux phases to better understand the process chemistry using a simpler mixture than sludge simulants.

---

[1] Processing limits are the same for both SRAT and SME as agitator and drive are identical.

Mercury Reduction and Stripping

Approximately 3.4 g of mercury were added to each simulation. The mercury recovery results are summarized in **Table** Error! No text of specified style in document.**-6**.

Table Error! No text of specified style in document.**-6. Supernate Testing with Mercury and Noble Metals**

Run	GF39a	GF39b	GF39c	GF39d	GF39e
Hg, wt %	1.5	1.5	1.5	1.5	1.5
Rh, wt %	0	0.0297	0.0297	0.0297	0
Pd, wt %	0	0.079	0.079	0.079	0
Ag, wt%	0	0.0014	0.0014	0.0014	0
Ru, wt %	0	0.3358	0	0.3358	0
% Koopman Acid Stoichiometry	100	100	100	80	100
% Hsu Acid Stoichiometry	74	74	74	60	58
Hg Collected, g	0.62*	None	0.43[#]	None	0.98

* Found 1.455 g of elemental Hg in kettle

[#] Found 1.939 g of black solids in kettle

The runs demonstrated that the mercury could be reduced and stripped with only glycolic acid (no formic acid). The exception to this is that in the runs with added ruthenium chloride (GF39b, GF39d), no mercury was recovered. Based on the obvious color changes (see photos below), the mercury was likely reduced in all the supernate runs. In runs with added ruthenium, 0.765 g Ru was added as $RuCl_3 \cdot 1.93H_2O$ (1.832 g or 0.0227 g-moles of Cl). In all runs, 3.689 g of HgO were added (0.0170 g-moles of Hg). In previous testing, the presence of Cl led to the production of calomel (Hg_2Cl_2), which is not steam stripped. It is recommended that these runs should be repeated with another form of Ru such as ruthenium oxide hydrate to see if adding the Ru without Cl has the same impact on mercury stripping.

The mercury (II) contained in the starting slurry as mercuric oxide was reduced during the glycolic acid addition at a pH of approximately 4.5. The photographs below (**Figure** Error! No text of specified style in document.**-2**) show the slurry both before and after the run from Run GF39a (mercury was added but no noble metals). The kettle contents quickly changed from the orange HgO slurry to a transparent silver colored solution over a period of several minutes. The silver color slowly disappeared during boiling when the mercury was being steam stripped and recovered in the MWWT.

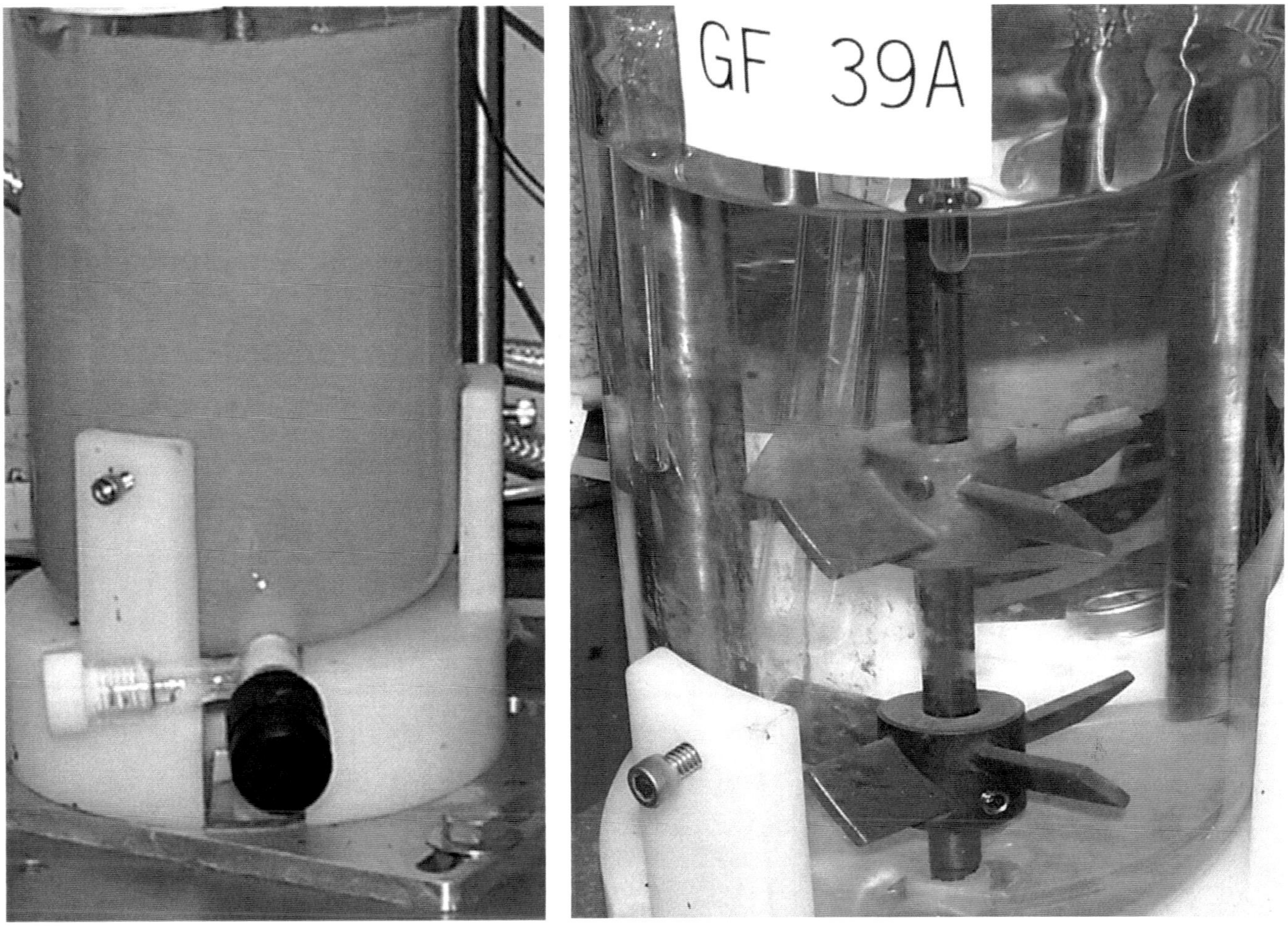

Figure Error! No text of specified style in document.**-2. Photographs of GF39a before and after SRAT cycle (Supernate plus HgO)**

In the runs with added noble metals and mercury, the slurry looked very much like sludge. The photographs below (**Figure** Error! No text of specified style in document.**-3**) show the slurry both before and after the run from Run GF39b (mercury and noble metals were added). The kettle contents quickly changed from the brown slurry to a transparent brown colored solution over a period of several minutes at a pH of 4.3. No mercury was recovered in the MWWT.

Figure Error! No text of specified style in document.**-3. Photographs of GF39b before and after SRAT cycle (Supernate plus HgO and noble metals)**

A mass balance was performed for each run to predict the concentration of all cations and anions throughout the run. In run GF39b (100% Koopman Stoichiometry, added noble metals and mercury), there was an apparent mass loss of 986 g (expected mass loss 151 g). This was calculated to match the final sodium concentration measured in the SRAT product sample. Using this mass loss, the predicted mercury concentration in the SRAT product is 2,306 mg/L and the measured mercury concentration was 2,315 mg/L. In other words, the mercury was completely soluble in the SRAT product and no mercury was recovered (not reduced, not stripped) in the MWWT. In contrast, run GF39a (100% Koopman stoichiometry, add mercury only), the final mercury concentration in the SRAT product was 14.9 mg/L compared to a predicted concentration of 1,433 mg/L (1.04% of the mercury was soluble). In addition, of the 3.4 g of mercury added initially on an elemental basis, 0.6 g was collected in the MWWT and 1.5 g was found in the SRAT product slurry as elemental mercury.

Nitrite and Carbonate Destruction

Nitrite and carbonate were below detection limits by the first hour of reflux in supernate testing. The results are summarized in **Table** Error! No text of specified style in document.**-7**.

Table Error! No text of specified style in document.**-7. Nitrite Data, mg/L**

Anion	GF39a	GF39b	GF39c	GF39d	GF39e
Post Nitric Acid	20,000	19,700	21,800	21,700	<500
Mid Glycolic Acid	10,800	6,560	4,540	11,700	<500
Post Glycolic Acid	<100	1,070	1,150	2,615	<500
1 hour dewater	<100	<100	<100	224	<500
Post Dewater	<100	<100	<100	<100	<500
Post Run	<100	<100	<100	<100	<500

Anion and Cation Mass Balance

Anions and cations were measured (solid lines in graphs below) throughout the supernate runs. A mass balance was completed for each run based on the known amounts added in preparing the supernate and the mass of added noble metals and mercury. These predictions (dotted lines), calculated by mass balance, were plotted along with the measured result in **Figure** Error! No text of specified style in document.**-4** and **Figure** Error! No text of specified style in document.**-5** (GF39b is presented as an example of this data). It should be noted that the PSAL measured nitrate agrees well with the nitrate prediction and the PSAL measured glycolate is approximately 20% higher than the prediction. In addition, oxalate is also much higher than predicted. It is likely that some oxalate is produced from glycolate decomposition. The measured nitrate is greater than predicted during glycolic acid addition due to the oxidation of nitrite to nitrate but is lower than predicted during reflux and boiling due to nitrate destruction. The sulfate concentration as measured by IC was very different than predicted. However, the measured sulfate, as calculated from ICP-AES S, was approximately 30% higher than predicted.

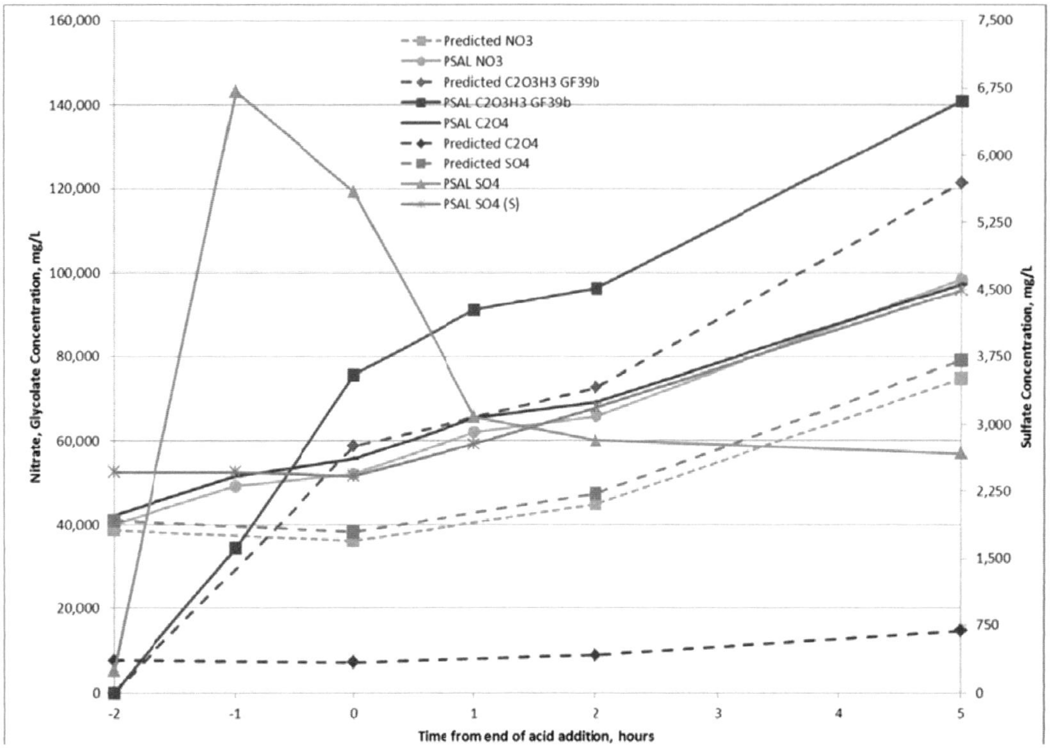

Figure Error! No text of specified style in document.**-4. Supernate Run GF39b Predicted and Measured Anion Concentration**

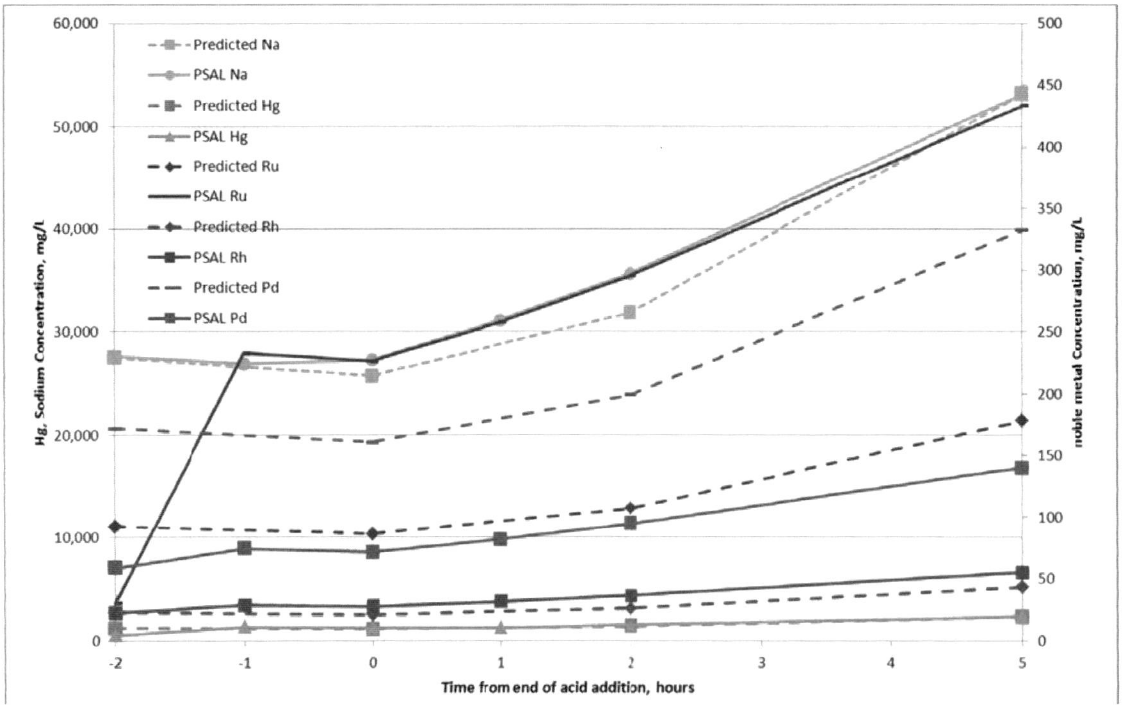

Figure Error! No text of specified style in document.**-5. Supernate Run GF39b Predicted and Measured Cation Concentration**

The measured sodium and mercury concentrations agreed well with predictions throughout run GF39b. The concentration of Pd and Ru were higher than predicted in Run GF39b. The Rh was approximately 80% of the predicted value and the Ag was below the detection limit. In Run GF39a (no added noble metals), the noble metals were all below detection limits.

Nitrite-free Supernate Test

One question that has always bothered our research team is whether glycolic acid is able to reduce mercury, allowing the elemental mercury to be removed by steam stripping. One theory was that glycolic acid needed a more reducing form such as glyoxylic acid to reduce the mercury. Glyoxylic acid, a better reducing agent than glycolic acid, could be produced by the reduction of glycolic acid by nitrite. As a result, a run was completed with a nitrite free simulant. This run was noble metal free, only mercury was added to the supernate. The result was that the mercury was completely reduced, a virtual duplication of Run GF39a. It appears that glycolic acid is fully capable of reducing mercury with or without nitrite present.

Slurry Testing

Twelve SRAT and six SME process simulations were completed to demonstrate the Glycolic-Nitric Flowsheet. Ten SRAT and four SME process simulations utilized the matrix sludges with added mercury and noble metals. Runs GF34, GF35, GF36 and GF37 were completed first and also included SME cycles. Runs 36b and 36c were duplicates of the GF36 SRAT cycle to compare the old and new processing rigs and determine whether the changes had impacted process chemistry. Run 37b was a duplicate of GF37 and GF38 was a higher acid stoichiometry repeat of GF37. The main reason for the four repeat runs was to better track mercury as the mercury recovery in the first four runs was poor. Runs 34b and 34c were duplicates of the GF34 SRAT cycle to determine whether lowering the purge impacted process chemistry. In addition, runs GF40 and GF41 were SRAT and SME process simulations designed to determine whether the Glycolic-Nitric Acid flowsheet could successfully process the less washed simulants. Some data from the supernate runs is included in this section for completeness if they were not reported in Section 3.1.

Off-Gas Hydrogen

A main objective of this testing was to show that hydrogen generation could be mitigated or eliminated by the use of the glycolic/nitric flowsheet. Hydrogen was detected only in the GF40 and 41 SRAT cycles. These two runs had the "new lower purge" which led to higher measured hydrogen concentrations for a given generation rate. The GC hydrogen quantitation limit is 0.005 volume %. The maximum hydrogen detected in these runs was 0.009 volume %, which would have been below detection limits with the current DWPF scaled purge. Note that these runs were completed at approximately 130% stoichiometry and produced approximately 1% of the hydrogen compared to essentially identical runs with the Baseline flowsheet. **Table** Error! No text of specified style in document.**-8** compares SRAT and SME cycle hydrogen on a DWPF scale. (**Figure** Error! No text of specified style in document.**-6**) summarizes the SRAT cycle hydrogen generation.

In the first four SME cycles (GF34, GF35, GF36, and GF37), formic acid was added with the frit in the SME cycle. In these runs, measurable hydrogen was generated, on the order of 0.05 volume percent. No formic acid was added in the GF40 and GF41 SME cycles. The GF41 hydrogen generation was just above quantitation limits in the GF40 SME cycle. In essentially identical runs with the Baseline flowsheet, the SME hydrogen limit was exceeded in runs at 125 and 130% acid stoichiometry.

Table Error! No text of specified style in document.**-8. Peak Hydrogen Generation**

Run	Sludge Composition	SRAT H2, lb/hr	SME H2, lb/hr
DWPF	Current Limit	0.65	0.223
GF34	HiFeHiMn	<0.0014	0.00556
GF34b	HiFeHiMn	<0.0014	No SME
GF34c	HiFeHiMn	<0.0014	No SME
GF35	SB7A	<0.0014	0.00398
GF36	HiFeLoMn	<0.0014	0.0111
GF37b	HiFeLoMn	<0.0014	No SME
GF37c	HiFeLoMn	<0.0014	No SME
GF37	LoFeLoMn	<0.0014	0.0157
GF37b	LoFeLoMn	<0.0014	No SME
GF38	LoFeLoMn	<0.0014	No SME
GF39a, b, c, d, e	Supernate	<0.0014	No SME
GF40	1.6 M Na	0.00287	0.00184
GF41	1.9 M Na	0.00324	<0.0012

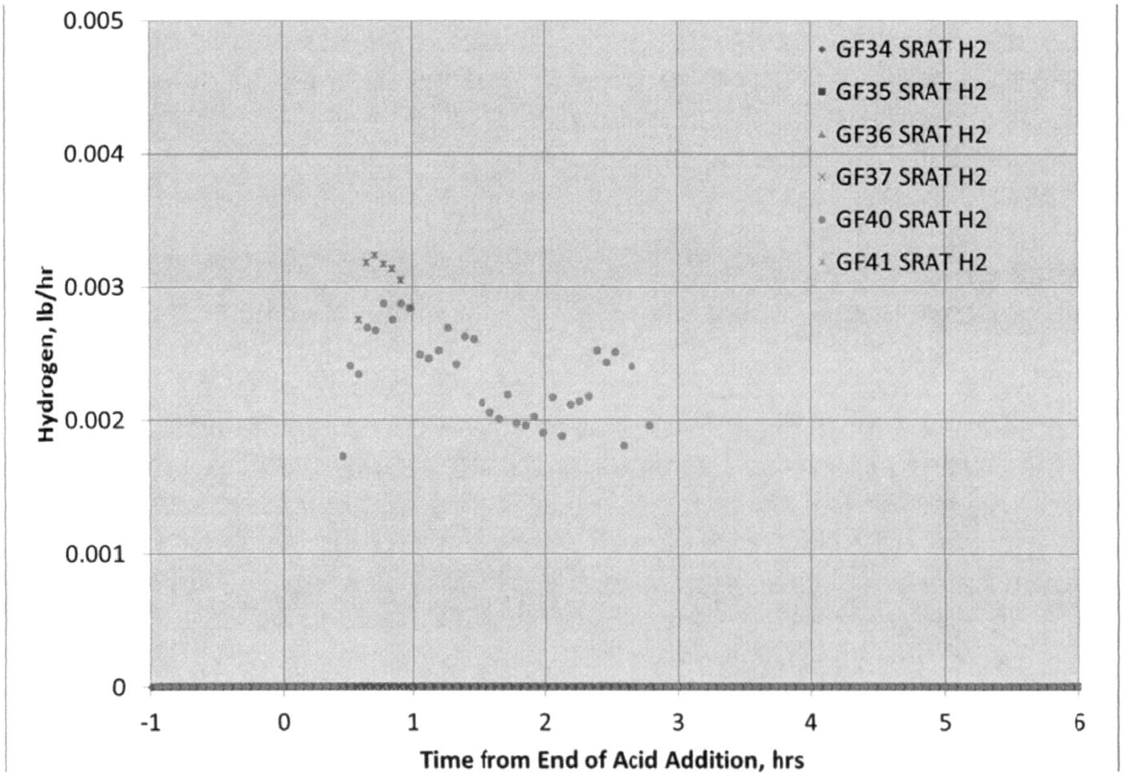

Figure Error! No text of specified style in document.**-6. SRAT Cycle Hydrogen Generation**

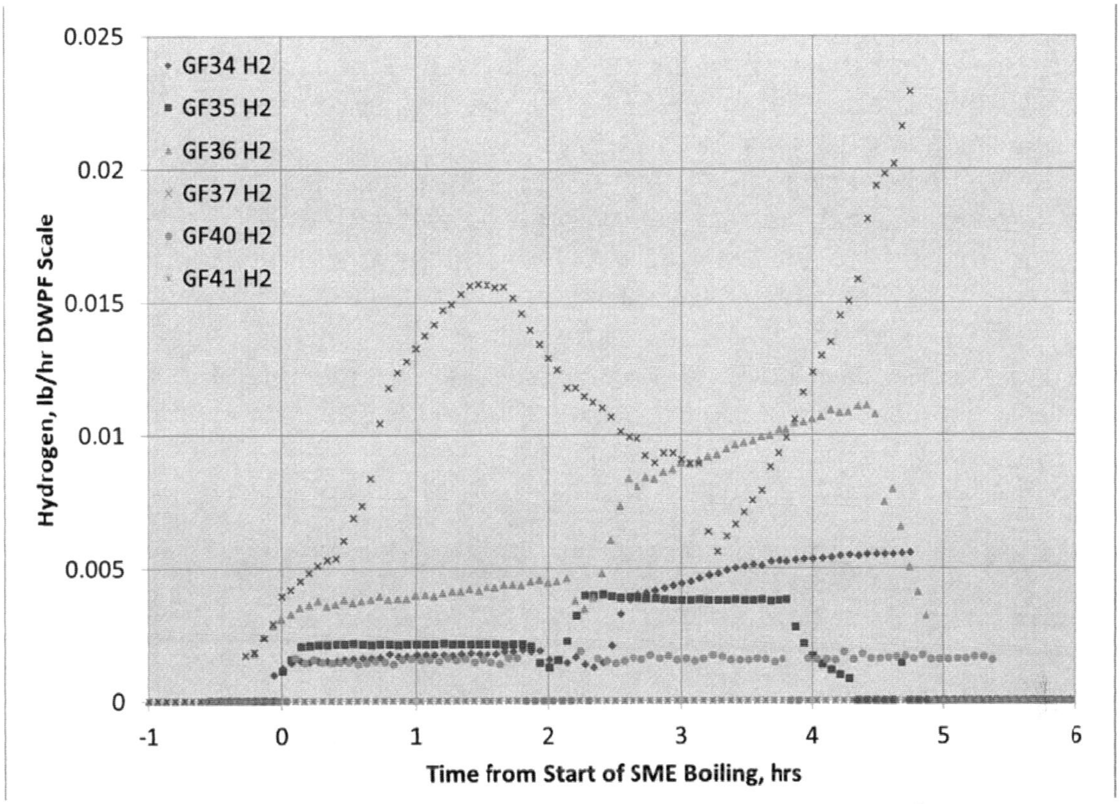

Figure Error! No text of specified style in document.**-7. SME Cycle Hydrogen Generation**

SRAT Mercury Reduction and Stripping

One of the most important questions to resolve concerning the glycolic-nitric acid flowsheet is whether mercury could be effectively reduced and steam stripped in the SRAT cycle without formic acid. In the Baseline flowsheet, mercury is reduced to elemental mercury during formic acid addition and then removed from the slurry by steam stripping during the concentration and reflux periods in SRAT processing.

The starting sludge was trimmed to 1.5 wt% Hg in the total solids. This required a theoretical boiling time of 12 hours to remove mercury to less than 0.80 wt% in the SRAT product total solids using lab-scaled DWPF design basis boil-up rates and a stripping efficiency of 750 g steam/g Hg.

A mass balance was completed for each of the runs to attempt to determine where the mercury had accumulated. The mercury mass balance is summarized in **Table** Error! No text of specified style in document.**-9**. In three of the first four runs, GF34, 35, 36 and 37, the mercury recovery was poor in the MWWT. As a result, Runs 34, 36 and 37 were repeated (Runs GF34b, GF34c, GF36b, GF36c, GF37b and GF38). The mercury recovery in the second set of runs was typical for lab-scale SRAT cycles[1]. No cause for the differences in duplicate runs has been

---

1 Zamecnik, J.R., Behavior of Mercury during DWPF Chemical Process Cell, SRNL-STI-2012-00051, REVISION 0, Savannah River National Laboratory, Aiken, SC, April 2012.

identified, but it is possible that there was technician error in collecting the mercury. Run GF35 (SB7A sludge) was not repeated, since the sludge was consumed in Run GF35. Run GF38 was performed at 125% acid stoichiometry to determine if acid stoichiometry impacted mercury recovery. Note that about 50% less mercury was recovered in the MWWT in run GF38 (125% acid stoichiometry) than was recovered in run GF37b (100% acid stoichiometry). This phenomenon is also seen in Baseline flowsheet runs.

Two SRAT and SME cycles (GF40 and GF41) were performed with two underwashed SB8 simulants to demonstrate that mercury can be reduced without formic acid in glycolic flowsheet runs with typical (not matrix) sludge simulant. These runs were completed in parallel with two Baseline flowsheet runs. The Glycolic-Nitric Flowsheet runs both had higher mercury recovery in the MWWT (same Koopman acid stoichiometry, same noble metals and mercury, essentially duplicate runs). In all Glycolic-Nitric acid flowsheet testing, the mercury stripping and recovery in the MWWT has either met or exceeded the recovery in the Baseline flowsheet runs.

Table Error! No text of specified style in document.**-9. Mercury Balance in SRAT and SME Cycle, g**

Run	% Acid Koopman	Added	MWWT	Slurry	Condensate	Total	% Recovery
GF34	104.0	10.56	2.27	12.12	2.27	16.66	160%
GF34b	104.0	10.56	5.89	6.36	NM	12.3	116%
GF34c	104.0	10.56	1.94	7.10	NM	9.04	86%
GF35	100.0	8.25	0.02	0.19	0.17	0.38	4.6%
GF36	106.1	10.17	0.14	5.85	0.53	6.52	64%
GF36b	106.1	10.17	2.27	6.59	NM	8.85	87%
GF36c	106.1	10.17	2.35	7.17	NM	9.53	94%
GF37	100.0	10.28	0.01	7.13	0.48	7.62	74.%
GF37b	100.0	10.28	4.10	4.75	NM	8.56	86%
GF38	125	10.28	1.99	7.40	NM	9.15	91%
GF40	130	11.15	3.79	5.21	0.44	9.44	85%
GF41	130	11.15	3.24	5.34	0.46	9.03	81%

Samples were taken periodically throughout the runs for mercury analysis. The chart below (Figure Error! No text of specified style in document.**-8**) shows the concentration of mercury in the slurry as a function of time for the eight runs. It is expected that the mercury concentration will decrease linearly during SRAT steam stripping and collect in the MWWT. A linear decrease of Hg concentration in the slurry assumes a constant boil-up rate and a constant approach to thermodynamic vapor-liquid equilibrium between the slurry and off-gas phases. The general trend of the mercury profile curves is a linear decrease as expected. It was expected that the SRAT product would have a mercury concentration of 0.8 wt% or 2160 mg/kg. The SRAT product Hg concentration ranged from 0.01-0.92 wt % total solids basis. Results are summarized **Table -10**.

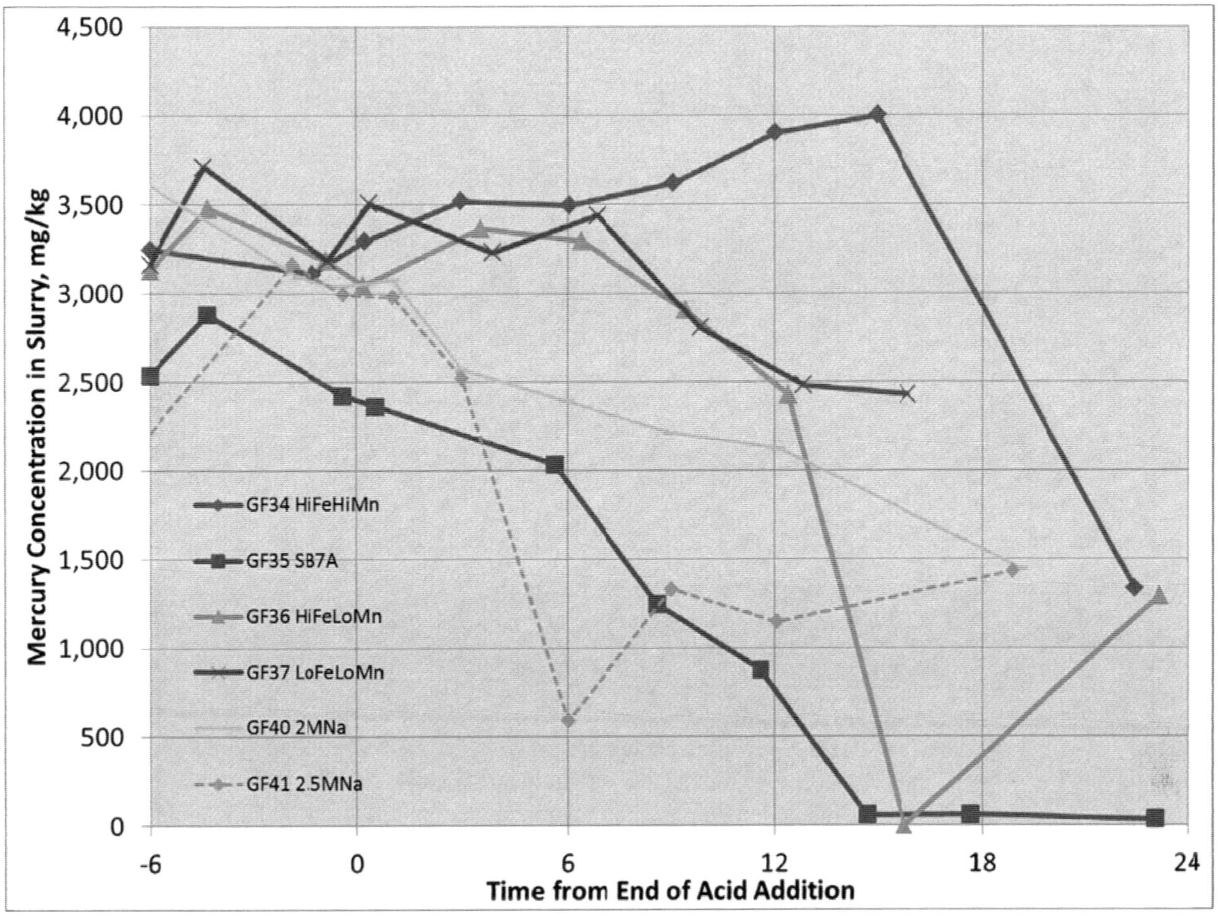

Figure Error! No text of specified style in document.**-8. Mercury concentration versus time in Selected SRAT and SME cycles**

Mercury is added to the sludge as HgO. In these runs the HgO was slurried with water and homogenized using the vortex mixer to break up any clumps and allow an even dispersal of the mercury. During SRAT processing the mercury is first dissolved and may later be reduced to elemental mercury. Once it is reduced, it is insoluble and can be steam stripped. In Runs GF37b and GF38, extra samples were pulled during the acid addition and dewater phase to understand when these reactions occur. In both runs, approximately 90% of the mercury was dissolved prior to the completion of nitric addition and the Hg was completely dissolved by midway through the glycolic acid addition. The mercury then is reduced during the first two hours of dewatering (faster during GF38, the 125% acid stoichiometry run, than during GF37b, the 100% acid stoichiometry run). The dissolution and reduction of mercury was very similar to that seen for Pd. The concentration of Hg and Pd are summarized in **Figure** Error! No text of specified style in document.**-9**.

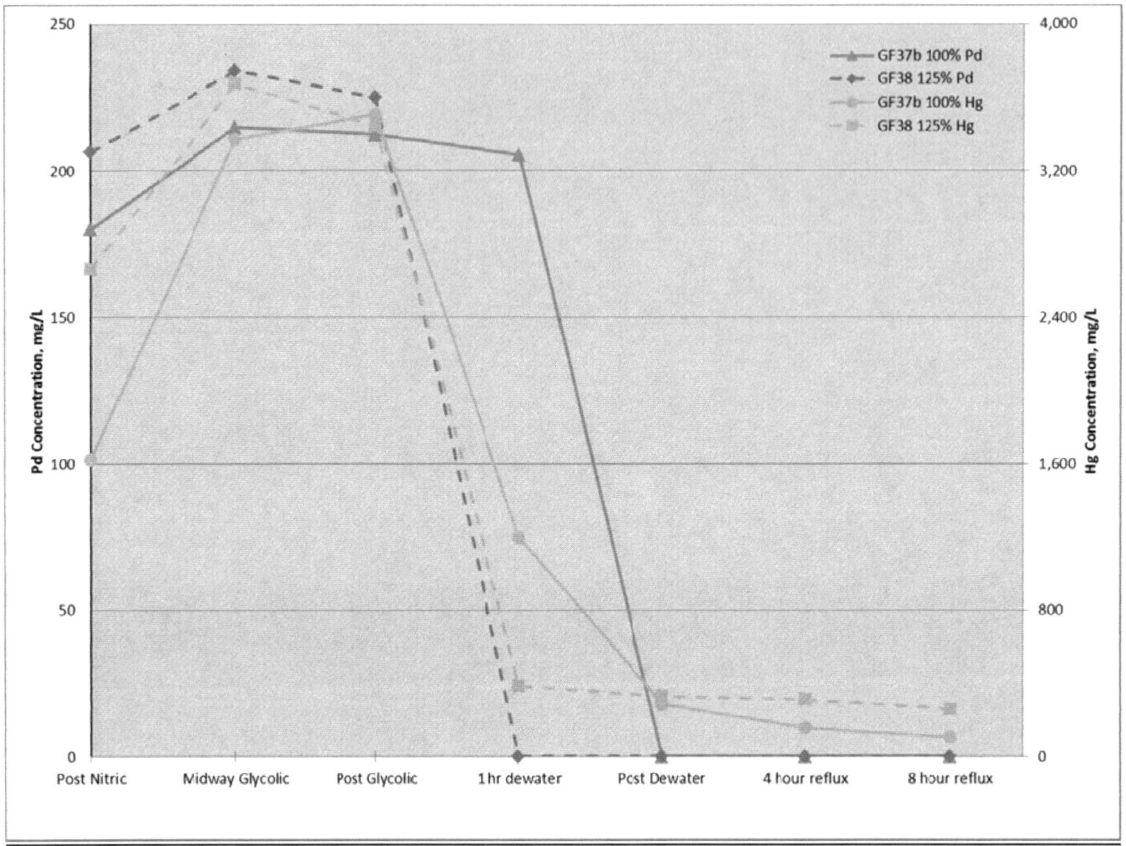

Figure Error! No text of specified style in document.**-9. Mercury and Palladium Concentration for GF37b and GF38 SRAT Cycles**

Mercury is being reduced by glycolic acid and approximately 30% of the mercury is removed from the SRAT by steam stripping and is collected in the MWWT. Approximately 40% of the mercury remains in the SME product. Another 4% was found in the condensate. There are two likely paths for condensate namely it can be removed by steam stripping (but not collect in the MWWT). This can happen in the SRAT cycle and SME cycle. In addition, the mercury can be dissolved by the strong acid in the condenser/MWWT condensate and overflow to the SMECT.

SRAT Elemental Data

General SRAT product slurry data for the twelve runs are tabulated below. Analyses were completed of both the slurry and supernate from all SRAT and SME products. The slurry results are summarized in **Table -10**. Conversion of the elemental data to the expected oxide form allows summing the oxides as a measure of both complete sample dissolution and accurate analysis of the major elements in the sludge product. The sum of oxides range from 98.2-100.5 over this data set (95-105 is considered acceptable). The slurry samples were filtered and the supernate results of these analyses are summarized in total solids basis

Table Error! No text of specified style in document.**-11**. The solubility of the cations is summarized in **Table** Error! No text of specified style in document.**-12**.

Table -10. SRAT Product Slurry PSAL Elemental Data, wt % calcined solids basis

Run	GF34	GF34b	GF34c	GF35	GF36	GF36b	GF36c	GF37	GF37b	GF38	GF40	GF41
Al	9.01	8.77	8.69	15.2	9.1	9.2	9.1	23.7	23.5	23.9	14.7	11.7
B	<0.100	NM	NM	<0.100	<0.100	<0.100	<0.100	<0.100	<0.100	<0.100	<0.100	<0.100
Ba	0.080	0.077	0.081	0.108	0.095	0.094	0.092	0.063	0.064	0.063	0.090	0.051
Ca	3.58	3.74	3.77	0.80	2.12	1.95	1.94	1.68	1.66	1.69	0.601	0.431
Cd	<0.010	NM	NM	<0.010	<0.010	<0.010	<0.010	<0.010	<0.010	<0.010	NM	NM
Cr	0.017	0.016	0.016	0.067	0.273	0.269	0.269	0.223	0.224	0.220	0.025	0.023
Cu	0.070	0.051	0.054	0.052	0.054	0.043	0.040	0.051	0.040	0.041	0.030	0.026
Fe	32.3	31.1	31.3	20.5	32.0	32.9	32.9	12.6	12.3	12.3	14.2	12.7
Hg@	1.27	0.82	0.89	0.02	0.65	0.81	0.87	0.83	0.59	0.92	0.52	0.52
K	0.080	0.087	0.118	0.075	0.061	0.077	0.083	0.071	0.079	0.086	0.412	0.442
Li	<0.100	NM	NM	<0.100	<0.100	NM	NM	<0.100	<0.100	<0.100	<0.100	<0.100
Mg	0.410	0.389	0.401	0.382	2.57	2.77	2.76	2.36	2.40	2.43	0.398	0.258
Mn	4.01	3.80	3.77	5.02	0.706	0.640	0.631	0.666	0.600	0.596	4.51	4.15
Na	14.2	13.8	13.9	15.5	13.5	13.4	13.2	15.3	14.1	14.4	23.0	26.1
Ni	0.212	0.182	0.189	3.42	2.69	2.73	2.73	2.37	2.35	2.37	1.95	1.55
P	<0.100	<0.100	<0.100	<0.100	<0.100	<0.100	<0.100	<0.100	<0.100	<0.100	<0.100	<0.100
Pb	0.080	0.069	0.071	0.023	0.056	0.049	0.041	0.058	0.050	0.039	<0.010	<0.010
Pd	<0.100	0.03	0.03	<0.100	<0.100	<0.010	<0.010	<0.100	<0.010	<0.010	<0.100	<0.100
Rh	0.031	<0.100	<0.100	0.032	0.033	<0.100	<0.100	0.047	<0.100	<0.100	<0.100	<0.100
Ru	0.032	<0.100	<0.100	0.031	0.032	<0.100	<0.100	0.030	<0.100	<0.100	<0.100	<0.100
S	0.276	0.282	0.281	0.347	0.276	0.264	0.269	0.294	0.283	0.260	0.292	0.341
Si	1.48	1.57	1.50	1.70	1.95	1.82	1.76	1.30	1.42	1.39	1.41	0.86
Sn	<0.010	NM	NM	0.029	0.107	0.102	0.102	0.089	0.094	0.093	NM	NM
Ti	0.010	<0.010	<0.010	0.025	0.010	<0.010	<0.010	0.010	<0.010	<0.010	0.022	0.012
Zn	0.062	0.063	0.065	0.064	0.072	0.071	0.070	0.061	0.062	0.062	0.046	0.039
Zr	0.055	0.051	0.054	0.236	0.117	0.110	0.110	0.045	0.040	0.044	0.201	0.178

@ Hg reported on a total solids basis

Table Error! No text of specified style in document.**-11. SRAT Product Supernate PSAL Elemental Data, mg/L supernate basis**

Run	GF34	GF34b	GF34c	GF35	GF36	GF36b	GF36c	GF37	GF37b	GF38	GF40	GF41
Al	292	1,060	1,230	217	411	2,280	2,250	554	2,210	4,040	1,720	1,970
B	1.22	NM	NM	1.28	1.23	<10.0	<10.0	1.49	<10.0	<10.0	<10.0	<10.0
Ba	2.18	4.33	4.44	0.99	1.65	3.28	3.32	1.26	2.69	3.37	9.37	11.0
Ca	2,390	4,150	4,200	109	2,390	3,350	3,490	2,150	3,040	2,870	368	403
Cd	<0.010	NM	NM	<0.010	<0.010	<0.100	<0.100	<0.010	<0.100	<0.100	NM	NM
Cr	2.58	4.34	4.85	3.52	32.0	54.8	53.7	86.0	103	198	27.4	29.8
Cu	11.6	20.8	23.0	1.33	11.9	25.9	24.6	15.1	24.7	38.5	19.4	21.9
Fe	1,670	4,470	5,220	141	1,040	2,810	3,290	328	1,490	3,560	5,510	4,950
K	392	303	422	321	272	16.4	5.23	290	265	247	918	985
La	27.8	NM	NM	2.84	18.6	239	269	39.2	NM	NM	<10.0	<10.0
Li	<10.0	NM	NM	<10.0	<10.0	<1.00	<1.00	<10.0	<1.00	<1.00	<10.0	<10.0
Mg	309	256	268	203	4,830	4,460	4,040	4,410	4,090	4,180	280	316
Mn	8,850	7,060	7,370	2,670	1,330	1,280	1,150	1,300	1,280	922	5,130	4,190
Na	30,100	24,900	15,400	33,700	26,500	28,200	25,900	30,400	29,800	28,500	44,300	47,800
Nd	7.13	NM	NM	0.61	4.58	10.77	9.84	10.71	NM	NM	NM	NM
Ni	121	117	123	100	2,940	3,160	2,850	3,160	3,180	3,960	732	659
P	0.86	<1.00	<1.00	1.04	1.13	1.54	1.86	1.56	<10.0	<10.0	14	20.6
Pb	4.10	4.73	5.14	0.17	0.61	2.29	2.01	2.31	4.15	18.9	6.73	9.91
Pd	0.16	0.123	0.122	0.21	0.16	0.18	0.13	0.18	<0.100	<0.100	<1.00	<1.00
Rh	2.97	15.9	10.4	9.21	10.7	18.4	15.0	12.0	36.0	78.3	30.5	31.4
Ru	106	174	125	25.6	181	229	206	289	330	453	54.1	65.9
S	645	439	446	880	572	453	482	672	599	513	583	622.7
Si	23.5	52.3	55.8	38.3	17.4	132	68.8	67.4	121	103	44.4	77.8
Sr	4.17	NM	NM	2.32	3.06	NM	NM	2.99	41.6	77.3	NM	NM
Ti	<0.010	<0.100	<0.100	<0.010	<0.010	<0.100	<0.100	<0.010	<0.100	<0.100	0.700	0.867
Zn	25.7	30.3	33.7	0.400	28.6	34.1	34.0	39.6	42.8	64.6	22.0	27.1
Zr	6.29	20.9	23.8	22.6	14.5	60.9	62.0	17.7	38.3	50.7	195	208

Table Error! No text of specified style in document.**-12. Major Components: SRAT Product % of Element Dissolved**

Run	Al	Fe	Na	Mg	Mn
GF34 HiFeHiMn	1.3	2.1	87.5	31.0	90.8
GF34b HiFeHiMn	0.0011	7.3	90.8	33.3	93.8
GF34c HiFeHiMn	0.0010	8.5	92.7	34.1	99.9
GF35SB7A	0.6	0.3	92.9	22.7	22.7
GF36 HiFeLoMn	2.0	1.5	88.2	84.0	84.6
GF36b HiFeLoMn	12.9	4.4	88.2	84.0	84.6
GF36c HiFeLoMn	12.7	5.1	108.6	83.2	103.4
GF37 LoFeLoMn	1.1	1.2	94.9	89.0	93.1
GF37b LoFeLoMn	4.8	6.3	108.8	88.1	110.2
GF38 LoFeLoMn	9.4	16.2	109.6	95.8	86.0
GF40	6.0	20.1	99.7	48.7	58.8
GF41	9.3	21.7	102	68.2	56.1

SRAT Anion Data

Ion Chromatography using weighted dilutions of samples (not the AD acid strike oxalate method) was performed on both the slurry and supernate from all SRAT and SME products. The slurry results are summarized in Table Error! No text of specified style in document.**-13**. The slurry samples were filtered and the supernate results of these analyses are summarized in Table Error! No text of specified style in document.**-14**. Anion balance data for nitrite, nitrate, formate and glycolate are presented in the table below for all runs (**Error! Reference source not found.**).

The SRAT and SME product oxalate results are of particular interest. The starting sludge contained about 800 mg/kg oxalate, which could be partially destroyed catalytically during the SRAT cycle. In the glycolic/formic flowsheet runs, however, oxalate was being created. The glycolic acid is likely oxidized to glyoxylic acid ($HCOCO_2H$) by nitrite, which is further oxidized to oxalic acid by the reduction of mercury. However, more experiments are needed to pinpoint the reaction pathways.

Table Error! No text of specified style in document.**-13. SRAT Product Slurry PSAL Anion Data, mg/kg Slurry Basis**

Run	Formate	Chloride	Nitrite	Nitrate	Sulfate	Oxalate	Glycolate
GF34	<100	650	<100	57,150	1,250	1,990	44,850
GF34b	<100	649	<100	54,450	1,910	4,970	46,450
GF34c	<100	717	<100	53,900	2,720	5,860	50,000
GF35	<100	572	<100	43,450	1,910	4,370	39,850
GF36	<100	622	<100	57,500	1,210	3,955	37,250
GF36b	<100	591	<100	56,650	1,280	3,190	51,250
GF36c	<100	602	<100	56,350	1,240	3,210	53,100
GF37	<100	821	<100	56,550	1,500	2,755	42,200
GF37b	<100	590	<100	52,500	1,445	2,420	55,450
GF38	<100	583	<100	56,900	1,420	2,655	77,850
GF40	<100	<500	<100	48,200	1,780	17,000	49,200
GF41	<100	<500	<100	41,700	1,220	13,300	48,600

Table Error! No text of specified style in document.**-14. SRAT Product Filtrate PSAL Anion Data, mg/L Supernate Basis**

Run	Formate	Chloride	Nitrite	Nitrate	Sulfate	Oxalate	Glycolate
GF34	<100	894	<100	80,500	2,250	1,570	56,300
GF34b	<100	806	<100	71,000	2,280	6,310	58,400
GF34c	<100	931	<100	70,900	3,520	7,730	66,500
GF35	<100	823	<100	63,700	2,790	3,800	48,600
GF36	<100	858	<100	86,300	2,170	3,250	46,700
GF36b	<100	736	<100	74,000	1,530	4,060	64,500
GF36c	<100	783	<100	74,300	1,610	4,240	70,700
GF37	<100	913	<100	82,100	2,740	3,860	61,300
GF37b	<100	772	<100	70,900	1,850	3,030	72,500
GF38	<100	746	<100	77,300	1,790	3,605	98,100
GF40	<500	<500	<500	60,400	1,910	11,100	50,300
GF41	<500	<500	<500	76,100	2,340	14,700	57,500

As a result of uncertainty of the anion analyses, four samples were submitted to AD for both TOC and anion analysis. The data below (**Table** Error! No text of specified style in document.**-16**) shows the results from both PSAL and AD for comparison. The agreement is fairly good, with the exception of the glycolate and oxalate. In addition, the carbon species (formate, oxalate, glycolate) were converted to carbon concentrations and summed to estimate the Total Organic Carbon (TOC) result for each sample. These results were compared to the AD measured TOC result. It is obvious that the TOC predicted from the PSAL results agreed well with the TOC measurement.

Table Error! No text of specified style in document.**-15. % Anion Dissolved in SRAT Products**

Run	Chloride	Glycolate	Nitrate	Oxalate	Sulfate	Sulfate (S)*
GF34	101.7	92.9	104.1	58.2	133.1	96.1
GF34b	99.6	59.3	98.7	41.6	114.8	78.5
GF34c	99.9	60.0	106.2	51.4	88.4	80.6
GF35	94.9	91.7	110.4	65.4	110.1	108.0
GF36	108.2	93.8	112.4	61.5	134.2	92.8
GF36b	98.5	97.9	101.6	99.0	92.9	88.9
GF36c	103.2	103.8	102.7	102.9	101.1	92.2
GF37	96.8	111.4	111.3	107.3	139.9	109.1
GF37b	101.3	101.8	105.2	97.7	99.4	109.6
GF38	104.8	98.3	105.9	105.8	98.5	109.6
GF40	85.2	60.1	94.1	61.1	103.6	103.4
GF41	101.9	65.5	94.4	93.8	157.2	101.7

* Sulfate (S) is a calculation of SO_4 from measured ICP-AES Sulfur analysis

Table Error! No text of specified style in document.**-16. SRAT Product AD and PSAL Anion with Comparison to AD TOC, mg/kg**

Analyte	GF36b	GF36c	GF37b	GF38
PSAL glycolate	50,200	55,100	55,500	77,900
AD Glycolate	33,900	34,400	35,900	54,500
PSAL Oxalate	3,160	3,300	1,340	2,390
AD Formate	<500	<500	<500	<500
PSAL Formate	<100	<100	<100	<100
PSAL Calculated TOC	20,900	24,100	24,100	32,400
AD Calculated TOC	11,500	11,500	11,900	18,100
AD Measured TOC	19,700	28,600	24,500	26,200

SME Elemental Data

General SME product sample data for the four runs (GF34, GF35, GF36 and GF37 had SME cycles) are tabulated below. The waste loading for these runs was targeted at 36% using frit 418. Elemental analyses were completed of both the slurry and supernate from all SME products. The slurry results are summarized in **Table** Error! No text of specified style in document.**-17**. Conversion of the elemental data to the expected oxide form allows summing the oxides as a measure of both complete sample dissolution and accurate analysis of the major elements in the sludge product. The sum of oxides range for 98.6-100.5 over this data set (95-105 is considered acceptable). The slurry samples were filtered and the supernate results of these analyses are summarized in **Table** Error! No text of specified style in document.**-18**.

Table Error! No text of specified style in document.**-17. SME Product Slurry Elemental Data, wt % calcined solids basis**

Run	GF34	GF35	GF36	GF37	GF40	GF41
Al	3.35	5.47	3.4	9.00	5.06	5.03
B	1.32	1.40	1.30	1.30	1.50	1.59
Ba	0.031	0.041	0.036	0.026	0.027	0.023
Ca	1.28	0.25	0.680	0.651	0.304	0.300
Cd	<0.010	<0.010	<0.010	<0.010	NM	NM
Cr	0.016	0.037	0.119	0.103	0.017	0.016
Cu	0.035	0.037	0.035	0.029	0.013	0.015
Fe	11.8	7.4	11.9	4.85	4.08	3.45
K	0.063	0.048	0.041	0.049	0.174	0.173
Li	2.20	2.31	2.27	2.18	NM	NM
Mg	0.152	0.145	0.924	0.858	0.115	0.102
Mn	1.44	1.79	0.231	0.220	1.35	1.13
Na	8.60	9.23	8.60	9.06	11.2	10.8
Ni	0.073	1.21	1.00	0.92	0.58	0.49
P	<0.100	<0.100	<0.100	<0.100	<0.100	<0.100
Pb	0.038	0.014	0.036	0.035	<0.100	<0.100
Pd	<0.100	<0.100	<0.100	<0.100	<0.100	<0.100
Rh	0.017	0.018	0.027	0.025	<0.100	<0.100
Ru	0.015	0.024	0.034	0.023	<0.100	<0.100
S	0.099	0.116	0.102	0.114	0.075	0.082
Si	23.35	24.3	23.45	22.9	25.0	26.2
Sn	<0.010	0.013	0.044	0.038	NM	NM
Ti	0.007	0.013	0.012	0.008	0.064	0.066
Zn	0.026	0.027	0.028	0.026	0.017	0.016
Zr	0.026	0.103	0.051	0.024	0.182	0.173

Table Error! No text of specified style in document.**-18. SME Product Supernate Elemental Data, mg/L supernate basis**

Run	GF34	GF35	GF36	GF37	GF40	GF41
Al	178	343	320	922	2,240	2,680
B	54. 0	55.0	48.0	47.2	<10.0	<10.0
Ba	2.22	1.06	1.68	1.35	14.3	16.5
Ca	2,090	169	2,110	1,960	399	419
Cd	<0.010	<0.010	<0.010	<0.010	NM	NM
Cr	1.93	7.03	34.6	69.4	35.3	37.7
Cu	5.82	3.16	12.8	16.7	27.0	28.8
Fe	1,280	326	1,200	973	10,100	12,250
K	396	311	252	212	991	1,095
La	18.2	10.2	17.6	36.6	NM	NM
Li	2670	216	234	183	352	376
Mg	315	223	4,660	3,460	5,230	4,680
Mn	8,610	3,620	1,280	998	46,300	52,800
Na	29,500	37,000	24,750	25,500	46,300	52,800
Ni	114	226	2,800	2,490	1,070	868
P	0.77	2.49	1.02	1.87	10.2	10.8
Pb	4.01	0.30	0.81	2.01	15.7	19.1
Pd	0.18	0.24	0.14	0.13	<1.00	<1.00
Rh	2.83	12.2	12.6	7.95	36.7	37.9
Ru	86.8	35.3	166	205	85	97
S	679	874	575	516	715	755
Si	30.6	102	27.6	71.5	62.3	86.7
Sr	4.22	2.67	3.05	2.77	NM	NM
Ti	<0.010	<0.010	<0.010	<0.010	0.938	1.194
Zn	20.6	2.40	28.4	17.0	41.9	44.2
Zr	5.25	38.0	32.1	22.2	261	258

SME Anion Data

Ion Chromatography was completed for both the slurry and supernate from all SME products. The slurry results are summarized in Table Error! No text of specified style in document.**-19**. The slurry samples were filtered and the supernate results of these analyses are summarized in Table Error! No text of specified style in document.**-20**. Anion balance data for nitrite, nitrate, formate and glycolate are presented in the table below for all runs (**Error! Reference source not found.**).

The anion data is inconsistent. For example, in Run GF37, the data indicates there was high nitrite to nitrate conversion in the SRAT and high nitrate loss in the SME. Also, it indicates that glycolate was destroyed in the SRAT and generated in the SME. It is more likely that there was a lower nitrite to nitrate conversion and lower glycolate loss in the SRAT with minimal nitrate and glycolate loss in the SME. The inconsistent results is likely due to fouling of the IC columns by metals and oxalate that are soluble at pH 4 but insoluble at pH 10 (approximately sample of eluent). It is recommended that removal of metals with an appropriate guard column be considered. An anion round robin has been initiated to resolve the issues with the analytical technique.

Table Error! No text of specified style in document.**-19. SME Product Slurry Anion Data, mg/kg Slurry Basis**

Run	GF34	GF35	GF36	GF37	GF40	GF41
Chloride	525	445	494	821	500	500
Nitrite	<100	<100	<100	<100	<100	<100
Nitrate	43,650	34,750	43,650	56,550	48,200	41,700
Sulfate	1,060	1,470	1,000	1,500	1,780	1,220
Oxalate	1,670	3,290	4,150	2,755	17,000	13,300
Glycolate	37,250	30,750	28,200	42,200	49,150	48,600
Formate	1,405	2,330	1,720	<100	<100	<100
Phosphate	<100	<100	<100	<100	<100	<100

Table Error! No text of specified style in document.**-20. SME Product Filtrate Anion Data, mg/L Supernate Basis**

Run	GF34	GF35	GF36	GF37	GF40	GF41
Formate	<100	<100	<100	<100	<500	<500
Chloride	892	770	884	726	<500	<500
Nitrite	<100	<100	<100	<100	<500	<500
Nitrate	79,100	58,200	84,750	60,100	94,000	84,000
Sulfate	2,345	2,825	2,710	2,570	2,330	2,540
Oxalate	1,620	4,845	3,960	3,395	10,400	13,500
Glycolate	59,400	42,150	45,550	45,600	60,400	58,500
Formate	1,715	3,785	2,251	1,940	<500	<500
Phosphate	<100	<100	<100	<100	<500	<500

SME Condensate

The SME condensate was not analyzed for GF34-38. However, GF40 and GF41 SME condensate samples were collected. Each sample was analyzed for elementals via ICP-AES, and anions via IC. The condensate was very low in anions and cations. The largest component is the silicon, likely an antifoam degradation product, not frit, as the same concentration was seen in the SRAT condensate. Note also that the pH of the SME condensate is considerably higher than the SRAT condensate. The SME dewater results are summarized in

Table Error! No text of specified style in document.**-21**.

Table Error! No text of specified style in document.**-21. SME Condensate, mg/L**

Analyte	GF40	GF41
Hg	24.6	15.4
K	14.6	14.5
Na	15.3	8.62
Si	1,455	553
Ti	<1.00	<1.00
NO_3^-	329	125
Density	0.9976	0.9970
pH	3.28	3.39

Note: The following were less than detection limits: Al, Ca, Cr, Cu, Fe, Mg, Mn, Ni, P, S, Ti, Zn, Zr, NO_2^-, SO_4^{2-}, $C_2O_4^{2-}$, $C_2H_3O_3^-$, HCO_2^-

Other SME Data

Other SME product data are summarized in the **Table** Error! No text of specified style in document.**-22**. Of particular note is that the GF37 SME was not completed prior to kettle breakage. As a result, the total solid result of the recovered product is significantly lower than had been targeted. Also, no analyses were completed on the SME condensate from Runs GF34-GF37.

Table Error! No text of specified style in document.**-22. Other SME Product Data**

Run	GF34	GF35	GF36	GF37	GF40	GF41
Total Solids, wt%	48.8	46.3	45.8	39.3	49.8	54.0
Insoluble Solids, wt%	37.9	37.6	35.3	29.7	36.9	41.0
Calcined Solids, wt%	38.5	37.3	35.8	29. 7	38.0	42.9
Soluble Solids, wt%	10.9	8.71	10.5	9.53	12.9	13.0
pH	4.66	6.18	4.39	4.31	4.76	4.81
Slurry Density, g/mL	1.42	1.34	1.38	1.29	1.137	1.150
Supernate Density, g/mL	1.11	1.10	1.10	1.09	1.243	1.234
Ammonium, mg/L	14	<5	<5	7	<10	<10

Note: GF40 and GF41 SME products were too thin and the frit settled quickly. It was difficult to maintain a uniform mixture. Higher total solid targets are recommended in future processing of underwashed sludges as significantly less insoluble solids are present in the SME product, as the sodium is included in the waste loading calculation.

SRAT Supernate Chemistry

The composition of the SRAT product slurry and supernate anions is summarized in Table Error! No text of specified style in document.**-13**, Table Error! No text of specified style in document.-14. SRAT Product **Filtrate PSAL Anion Data, mg/L Supernate Basis** and **Error! Reference source not found.**. Figure Error! No text of specified style in document.**-10** and Figure Error! No text of specified style in document.**-11** show the amount of each element found in the SRAT product supernate expressed as a percentage of the total element present. These data are calculated by dividing the supernate concentration (converted to mg/kg on a slurry basis) by the total slurry fraction of each element (converted to mg/kg). Numbers greater than 100% are not physically possible and are a result of error in one of the analytical measurements used in the calculation.

The % solubility of each anion is approximately 100% (80-120% based on method uncertainty), except for oxalate, which had a solubility of approximately 60% in the GF34 and GF35 runs. The solubility of Al and Fe was low in all runs. The solubility of Na, Mg, and Mn are all high in the glycolic flowsheet runs. For most of the metals, which are present primarily as hydroxides and oxides in the sludge, the concentration in the supernate increases throughout the SRAT cycle, but appear to be constant by the end of the SRAT cycle. Samples were pulled at the completion of nitric acid addition, midway through glycolic acid addition, after completing glycolic acid addition, one-hour into dewater, post dewater, 4 hours into reflux and 8 hours into reflux. These samples were centrifuged soon after being pulled to make sure no further reactions occurred due to insoluble solids. One interesting observation is that the centrifuged GF37b samples (100% acid stoichiometry) had almost no supernate after centrifuging at the completion of dewater (0.3 g of supernate typical in these samples). Prior to dewater and throughout run GF38 approximately 6-7 g of supernate was easily removed after centrifuging.

Based on this data, the order of dissolution for the "major components" is: Hg>Ca>Mn>Ni>Mg>Al>Fe. The data is summarized for major metals (>1,000 g/L) in **Figure** Error! No text of specified style in document.**-10** and minor metals in **Figure** Error! No text of specified style in document.**-11**. The graphs show the approach to maximum solubility, defined as 100% for each element on these graphs. This does not indicate that 100% of these individual elements went into solution during processing.

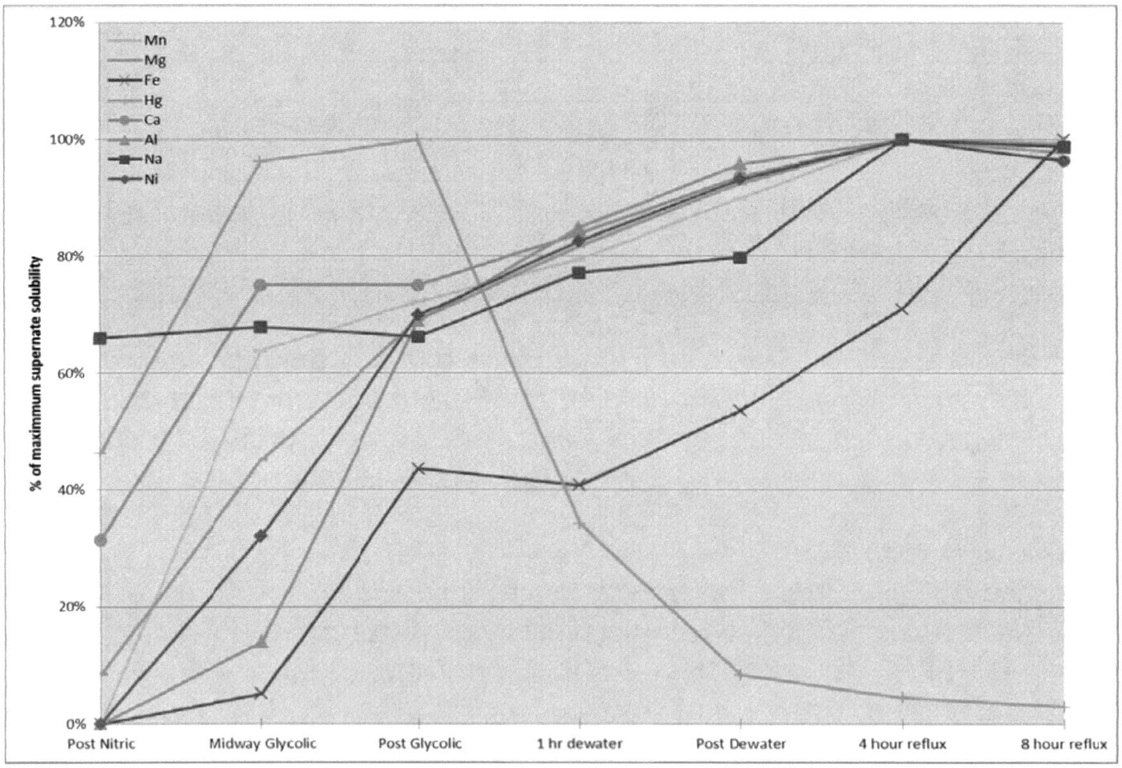

Figure Error! No text of specified style in document.**-10. Order of Dissolution of "Major Metals" During SRAT Processing**

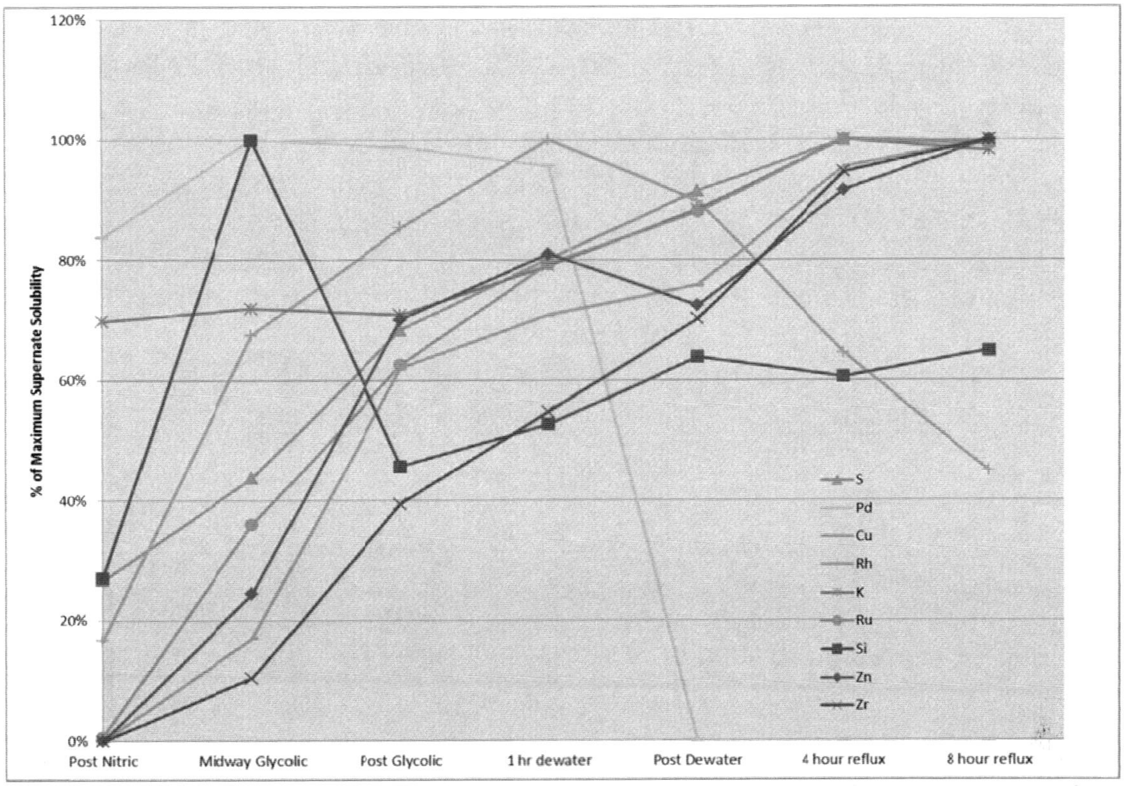

Figure Error! No text of specified style in document.**-11. Order of Dissolution of "Minor Metals" During SRAT Processing**

Several metals are of particular interest during SRAT processing. Note that mercury is discussed in Section 0. The reduction of Mn is important especially in the melter and cold cap in order to minimizing foaming in the melter. As can be seen in Figure Error! No text of specified style in document.**-10**, the Mn is dissolved (and likely reduced) early in the SRAT cycle and is >90% of the maximum solubility by the end of dewater. There are several metals that are essentially totally soluble such as Na, K, and Ca. The concentration of each metal changes as the metal is first diluted during acid addition, then concentrated during the dewater phase. In addition, the concentration of soluble metals should remain constant throughout the post dewater stage of the SRAT cycle. In GF37b, the concentration of both Na and Ca increased during this time, likely due to the extended centrifuge time necessary to squeeze out the 0.3 g of supernate from a 15 mL centrifuge tube. Note that for the lower acid run, GF37b, the calcium was not completely soluble until midway through glycolic acid. This may indicate that it may take more than 100% acid stoichiometry to produce a SRAT product that is easily concentrated in the SME.

In order to understand the dissolution of metals and the timing of their dissolution, additional samples were pulled during runs GF37b and GF38 (LoFeLoMn sludge). The dissolution of Hg is discussed in the mercury section.

SME Supernate Chemistry

The main change in supernate chemistry during the SME cycle is that formic acid is added to the frit slurry to prevent caking. Formic acid is very reactive in DWPF SRAT and SME processing, ultimately leading to the noble metal catalyzed decomposition to hydrogen and CO_2. No formic acid was added or detected during the SRAT cycle. The solubility of the anions during the SME cycle is summarized in **Table** Error! No text of specified style in document.**-23**.

Table Error! No text of specified style in document.**-23. Major Components: SME Product % of Anion Soluble**

Run	GF34	GF35	GF36	GF37	GF40	GF41
Formate	68.1	92.5	76.7	67.4	NA	NA
Glycolate	89.0	78.1	94.6	95.8	62.4	57.6
Nitrate	101	95.4	114	101	99.0	96.5
Oxalate	54.1	83.9	55.9	77.4	31.1	48.6
Sulfate	123	109	159	136	66.4	99.7
Sulfate (S)	98.9	115.0	91.9	98.4	128	103

* Sulfate (S) is a calculation of SO_4 from measured ICP-AES Sulfur analysis

Conclusions

Testing was completed to demonstrate the viability of the newly developed glycolic/nitric flowsheet for processing in the Defense Waste Processing Facility's (DWPF) Chemical Process Cell (CPC). The Savannah River National Laboratory (SRNL) initiated a sludge matrix study to evaluate the impact on CPC processing. Four sludge simulants were designed to cover a broad insoluble solid composition range to bracket future sludge batches. The first pair of sludge parameters was high iron/low aluminum versus low iron/high aluminum (referred to as HiFe or LoFe in this report). The second pair of sludge parameters was high calcium-manganese/low nickel, chromium, and magnesium versus low calcium-manganese/high nickel, chromium, and magnesium (referred to as HiMn or LoMn in this report). In addition, a simple supernate simulant was prepared to match the composition of the matrix simulants.

Ten experiments (GF34 to GF37 and GF34b, GF34c, GF36b, GF36c, GF37b and GF38) were completed to demonstrate the glycolic-nitric flowsheet viability using the sludge matrix simulants. In addition, two experiments were performed with less washed simulants (GF40, 2M and GF41, 2.5 M Na endpoints) to demonstrate the viability of processing these sludges. Also, five supernate experiments (GF39a-GF39eGF39e) were performed to better understand the reaction sequence, particularly the reduction and stripping of mercury.

Composition and physical property measurements were made on the Sludge Receipt and Adjustment Tank (SRAT) and Slurry Mix Evaporator (SME) products. Composition measurements were made on the composited condensates from the Mercury Water Wash Tank (MWWT), and Formic Acid Vent Condenser (FAVC), on the ammonia scrubber solution, and on SRAT samples pulled throughout the SRAT cycle. Updated values for glycolate and formate loss, nitrite-to-nitrate conversion, and oxalate formation were found that can be used in the acid calculations for future process simulations with the glycolic-nitric flowsheet.

Preliminary results of the initial testing indicate:

- Hydrogen generation rate was below detection limits (<11.4E-3 lb/hr DWPF-scale or <0.005 vol%) throughout all SRAT cycles with matrix simulants. Hydrogen generation rate was above detection limits for the less washed simulants (3.2E-3 lb/hr DWPF-scale or 0.009 vol%) due to the higher acid stoichiometry and the lower offgas purge.
- Hydrogen generation rate was below 0.0258 lb/hr DWPF-scale throughout all SME cycles with matrix simulants. Hydrogen was produced in the matrix SME cycles because formic acid was added with the frit slurry. Hydrogen generation rate was above detection limits for the less washed simulant in GF40 (1.8E-3 lb/hr DWPF-scale or 0.007 vol%) but was below detection limit in GF41 due to the higher acid stoichiometry and the lower offgas purge. No formic acid was added in runs GF40 and GF41.
- Mercury was both reduced and stripped without formic acid. The mercury concentration of the SRAT product was below the 0.8 wt % limit in eight of the runs and below 0.92 wt % in the other four runs.
- Nitrite in the SRAT product was <100 mg/kg slurry for all runs.
- Foaminess was not an issue using the nominal antifoam addition strategy or with reduced antifoam in these tests.
- High wt % total solids were achieved while staying within rheological limits which makes the glycolic acid/nitric acid flowsheet an improvement for processing more viscous sludges. However, there may be a tradeoff between excessive dissolution of metals and thinner rheology.
- The pH remained steady throughout processing (i.e. no pH rebound) potentially leading to more consistent processing during the CPC. The SRAT and SME products pH varied from 3.5-5.0 for the 100% and 130% acid stoichiometry runs, significantly lower than is typical of the Baseline nitric acid/formic acid flowsheet.
- The testing apparatus has been significantly modified to improve processing with high viscosity slurries. Testing of the old style and new style rig identified no differences in CPC processing, including steam stripping of Hg.
- The SRAT lower air purge was demonstrated in Run GF34c and used in GF40 and GF41. The SRAT purge can be reduced from 190 scfm to 93.7 scfm without negatively impacting DWPF CPC processing.
- Runs GF40 and 41 demonstrated that processing of less washed sludges is viable with the Glycolic-Nitric flowsheet. However, this flowsheet has not been demonstrated with ARP, MCU or actual waste.
- Several processing improvements were demonstrated in these runs including adding acid during heat-up, adding both acids at higher volumetric flowrates than are currently used in DWPF, and concentrating the SRAT during acid addition. Each of these improvements has the potential to shorten CPC processing time.

Recommendations

The glycolic-nitric flowsheet is recommended as a viable flowsheet alternative to the Baseline DWPF flowsheet. In the testing that has been performed to date, this flowsheet meets or outperforms the current flowsheet in minimizing off-gas generation, removing mercury, and producing a rheologically thinner product. Previous testing with glycolic/formic acid mixtures demonstrated a wide processing window regarding both the glycolic-formic ratio and acid stoichiometry. The addition of glycolic acid leads to SRAT products that are rheologically less viscous which means that more concentrated products can be produced, leading to potentially higher waste throughput per batch. In addition, the combination of lower pH processing and the complexing power of glycolic acid leads to the dissolution of more metals, which may minimize deposits in the CPC processing vessels and prevent the fouling of steam coils. Follow-up testing is recommended in the following areas:

- Improve glycolate and oxalate analyses. The majority of the glycolate results reported were correct. However, there are issues with anion and cation deposition on the column of the Ion Chromatograph (IC), causing higher than expected glycolate and oxalate in blanks and some samples. Both Process Science and Analytical Laboratory (PSAL) and Analytical Development (AD) have reported results that have varied significantly from expectations. Modification to the sample preparation method is likely needed to improve analytical accuracy and minimize the cleaning and replacement of the IC column. An alternative to the IC measurement of glycolate should also be considered.
- Determine the appropriate REDOX model for the glycolic-nitric flowsheet. The REDOX model may need more terms due to the more extensive reduction of some metals, including Mn and Fe. In addition, accurate measurement of glycolate (and possibly oxalate) and nitrate is needed to accurately predict REDOX. REDOX testing of the matrix sludges should be repeated using acceptable frits that meet Product Composition Control System (PCCS).
- Testing should be completed with alternate forms of ruthenium to determine whether the elimination of the chloride added as ruthenium chloride would improve the reduction and stripping of the mercury. Comparison testing should be completed with the Baseline and glycolic-nitric flowsheets.
- Test the glycolic-nitric flowsheet at acid stoichiometries of less than 100%. Demonstration of this flowsheet at an acid stoichiometry of <100% is recommended and might be useful for mercury stripping.
- Demonstrate the glycolic-nitric flowsheet (previously demonstrated in SRAT cycle with 80:20 glycolic:formic acid blend) with actual waste in SRNL Shielded Cells SRAT and SME processing, to include periodic slurry sampling throughout the SRAT and SME processing along with a glass REDOX measurement.
- Add the nitric and glycolic acid flowrate at the same scaled molar flowrate as formic acid to minimize glycolic-nitric flowsheet batch time.

- The nitric acid can be added during heat-up to decrease the SRAT cycle time. The nitric acid primarily neutralizes soluble the base species in the slurry with little offgas generation.
- Improve understanding of process chemistry, the decomposition of glycolate and the production of oxalate which are important to REDOX.
- Improve understanding of mercury reduction, stripping and accumulation during processing. Determine whether alternative equipment or processing changes are needed to maximize the collection of mercury in the Mercury Water Wash Tank.
- If confirmed by actual waste testing and larger scale testing with simulants, the antifoam addition can be reduced for this flowsheet. The addition of 100 mg/kg prior to glycolic acid addition, 100 mg/kg prior to boiling and 100 mg/kg each 12 hours of processing was adequate during simulant testing.
- More rigorous data collection is needed to validate the OLI aqueous model's solubility predictions with sample results. The methodology is summarized in the discussion.

(738f) Rheological Modification On Nuclear Waste Simulant Slurries: Correlation Between Physicochemical Characteristics and Rheological Modification

Jaehun Chun and Paul R. Bredt, Pacific Northwest National Laboratory, Richland, WA

Understanding and controlling rheological properties of slurries has been one of the most challenging problems in many industries. For nuclear waste treatment at the Hanford and Savannah River sites, it is extremely challenging due to the wide distribution of particle size, chemical species, and mass/volume fraction of particles. Operation of the waste treatment facilities at increased solids loading reduces the evaporative load on the melter systems and thereby increases waste processing rates. However, a higher solids loading creates significant processing issues associated with rheological properties. Controlling rheological properties allows for sustainable operations at higher solids loading and therefore reduces the operational cost and increases the waste processing rates. The use of several rheological modifiers over different nuclear waste simulant slurries has been investigated to address this issue. Using rheological analysis, it was found that weak organic acid type rheological modifiers (e.g., citric and polyacrylic acids) are generally good rheological modifiers. More importantly, the performance of rheological modifiers was found to be dependent on a coupled interplay between physicochemical parameters. More in-depth physical insights on the rheological modification will be discussed.

(738g) Progress in Removal of Radioactive Waste From Light Water Reactors

Lev Reznikov[1], Luke Sicard II[2] and Victor Zents[3], (1)Engineering, Global Fabrication, Inc., DuBois, PA, (2)Engineering, Global Fabrication Inc., DuBois, PA, (3)Global Fabrication Inc., DuBois, PA

There are currently in excess of one-hundred (100) light water reactor nuclear power plants operating within the United States; and many others in foreign countries. Radioactive corrosion products are present in varying accumulations throughout all light water reactor plants. The radioactive corrosion products/deposits are commonly referred to as "CRUD" an acronym for Clinch River Unidentified Deposits, a term coined to describe the usually black or gray, highly radioactive deposits that were first noted in the very early experimental light water reactor plants; most notable at the Clinch River facility.

The presence of CRUD inside the station piping system drives up the cost of routine operations and maintenance of the plant many fold. Simple, inexpensive tasks become highly intricate and inordinately expensive because of the extreme methods that must be employed to cope with the radioactivity emanating from the CRUD, and to prevent the release of CRUD to the stations environs.

Current filtration technology can remove the CRUD from the station fluid systems; however, collection of significant quantities of CRUD in conventional filters creates major radiological problems in removing the spent cartridges. The radiological penalties associated with cartridge change-out are nearly as bad as those associated with leaving the CRUD in place, inside the fluid system.

The radiological problems associated with cartridge change-out have been so troublesome that they have severely limited the use of cartridge filtration technology in commercial nuclear power plants. Finding a way to remove CRUD from troublesome areas without creating additional radiological problems in the process would be viewed as a noble effort, indeed. That is precisely what the new APPARATUS FOR RADIOACTIVE PARTICULATE FILTRATION filter/cartridge change-out system does. It will allow unimpeded utilization of the unique filtration technology without the severe radiological penalties associated with current cartridge change-out systems.

Our patented process of filter change-out allows us to remove the dirty filter cartridges without having to disassemble the vessel for cleaning. This in turn will significantly reduce personnel exposure as well as overall filter change out time.

(738h) Development of Nuclear Waste Tank Sludge Removal Using Scale Model

John L. Steimke, Savannah River National Laboratory, Aiken, SC

Hanford Site is in the process of removing radioactive waste from large storage tanks. The tanks contain salty supernate, salt and sludge. Nearly all of the actinides, such as uranium and plutonium, are in the sludge layer on the bottom of the tank. Estimates of plutonium inventories indicate a potential criticality concern in some tanks. Waste removal should not be performed in a way that significantly concentrates dense plutonium particles in the heel. Savannah River National Laboratory (SRNL) developed a four component sludge simulant, including a surrogate for plutonium particles, which has rheology, density and settling properties typical of Hanford tank waste. SRNL tested sludge removal in a 1:22 scale model waste tank using rotating submerged jets. Photographic and laser techniques were used to measure the shape and volume of the residual tank heel. Magnetic separation and a colorimetric technique were used to measure compositions of the heel and also of solids removed from the tank. Sludge removal efficiency and tendency to concentrate the plutonium surrogate were measured. The results will be used to guide larger scale tests.

AUTHOR INDEX

AUTHOR INDEX